FISHES
Aid to Collection, Preservation and Identification

FISHES
Aid to Collection, Preservation and Identification

By
Professor M.K. Jyoti
Head
Department of Zoology,
University of Jammu, Jammu
&
Dr. Arti Sharma
Field Biologist
Department of Zoology,
University of Jammu, Jammu

2016
Daya Publishing House®
A Division of
Astral International Pvt. Ltd.
New Delhi - 110 002

First Published, 2006
Reprinted, 2016

ISBN:978-93-5130-982-6 (Hardbound)

Publisher's note:

Every possible effort has been made to ensure that the information contained in this book is accurate at the time of going to press, and the publisher and author cannot accept responsibility for any errors or omissions, however caused. No responsibility for loss or damage occasioned to any person acting, or refraining from action, as a result of the material in this publication can be accepted by the editor, the publisher or the author. The Publisher is not associated with any product or vendor mentioned in the book. The contents of this work are intended to further general scientific research, understanding and discussion only. Readers should consult with a specialist where appropriate.

Every effort has been made to trace the owners of copyright material used in this book, if any. The author and the publisher will be grateful for any omission brought to their notice for acknowledgement in the future editions of the book.

Published by : **Daya Publishing House®**
A Division of
Astral International Pvt. Ltd.
– ISO 9001:2008 Certified Company –
4760-61/23, Ansari Road, Darya Ganj
New Delhi-110 002
Ph. 011-43549197, 23278134
E-mail: info@astralint.com
Website: www.astralint.com

Laser Typesetting : **Classic Computer Services**
Delhi - 110 035

Printed at : **Replika Press Pvt. Ltd.**

Acknowledgement

The students of fishery remain indebted to persons like Late Sh. Sain Dass (Fisherman), who assisted research scholars of this department for about a decade and many more of the kind whose experience and field knowledge of capture, identification, local nomenclature, measurements and behaviour of fish in nature became a source of enlightment, inspiration and encouragement for fishery students in the field.

Authors

Foreword

The book **"Fishes : Aid to Collection, Preservation and Identification"** is a valuable addition to the treatises on understanding the study of fish and fisheries. This treatise is authored by an experience teacher whose single-minded devotion and understanding of the subject is reflected appropriately in the concept and scheme of the book. Therefore, the outcome is very useful not only for beginners in the subject but also for those practising Ichthyology both in field and laboratory.

The challenge posed by Fishery Science lies in the numerical strength of species, diversity in morphological variations, variation in their habitat and adaptations thereof. Their habitat varies from 7 miles below to 3 miles above sea level. Their adaptation to the environmental regime in terms of habitat variations has been dealt within this book, fairly satisfactorily.

The book fulfills a void in the hitherto available information of different methods of fish catching, particularly in vogue in freshwaters of India. It carries useful information for fish consumer by indicating the index of freshness of edible fishes, generally marketed sometime after a gap of time of the catch.

For researchers, museologists and students of Ichthyology, pursuing practical courses, besides fish catching methods, information on fish preservation and identification of fishes is a welcome information.

Fins of fishes also exhibit interesting variations, which have been described in sufficient details with appropriate illustrations for the benefit and understanding of readers.

The chapter on collection of eggs, juveniles and larvae of fishes is another unique treatment in the book. Many authors have written information related to adult fishes, but very little importance has been accorded in the past to study of eggs, larvae and juveniles (fry and fingerlings), their habitat and identification. This book is different from previous works in as much as that this book gives due importance to the study of juveniles, which is an important area for fishing prospects.

Professor Y.R. Malhotra
Ex-Vice Chancellor, Jammu University
Ex-Member, Jammu & Kashmir Legislative Council

Preface

We feel pleasure in presenting this book, **"Fishes : Aid to Collection, Preservation and Identification"**, to a mixed class of readers who may be students of Ichthyology, Fishery or those working with commercial organisations for fish export or one's interested in fish cooking.

Thus, while developing the book contents, a special care has been taken to make it palatable by diverse sections of society, professional, non-professional and consumers. The stimulus for writing a book of this kind has come through our interactions with students of Zoology and Fishery Science. Often it was felt that students feel handicapped when confronted with certain terminologies which have escaped their vision and no consolidated literature is readily available to bridge continuity of understanding at occasions most needed. Break, thus generated, detracts and delinks attention from the real approach of understanding and gaps hardly get filled subsequently. Similarly most of the fish sellers and buyers sell and purchase fish without really knowing their real, local or scientific names. Hence an effort has been made to let them understand fish through appearance and diversities in fish morphology as has been cited in this book through supporting pictures (line drawings) for different variables in structures and terminologies so often used for them in India or other countries of the world.

The contents of this book are developed from the existing literature (books or research publications) and pictures drawn, modified or borrowed occasionally, realising the purpose for which these are being introduced to reader.

It is, therefore, a pleasure to thank each and every author whose work we have cited as well as those who might have not been referred inadvertently.

We shall feel glad to receive constructive suggestions from students/teachers and all those who read this book and feel concerned for its further improvement.

M.K. Jyoti and Arti Sharma

Contents

Chapter 1
Introduction

Fishes are cold-blooded aquatic vertebrates that possess paired and unpaired fins and breathe through gills. They seem to have acquired mastery of living in water through adaptations to all operative conditions from deep sea to snow-melt torrential streams, from desert pools to cave waters, from sub-terrarian waters to coral reef or from frozen waters of Antarctica to hot springs. In fact they occupy all possible habitat from 3 miles (4.8 km) above sea level upto 7 miles (11.2 km) deep in the sea (*Figure 1*).

Fish has been enjoying a very special consideration and place in human civilisation from times immemorial and occupies an important place in Indian mythology, history and traditions. One of the incarnations of God in Indian mythology was in the form of *"Matsyavathara"*. Interest in fish eating dates back to the dawn of history. It is believed that hunting of fish was not uncommon in the pre-historic times and was prevalent in stone-age. Fish remains, indicative of their use as food by ancient people, were obtained in excavations at Mohenjodero and Harappa of Indus-Valley Civilisation (2500 BC–15 BC). The representational continuity of fish in India is traceable from the pottery of Nal and the Indus Civilisation (Jhingran, 1991).

The information on fishing during the palaelethic age is rather sparce and not much reliable. It seems fishing was already practised by the pre-homonid ancestors of man. The Neanderthal man, about 10,000 years, had a certain preference for waters where trout and other fishes could be caught by hands with the help of stone barriers in rivers.

During the earliest stages of civilisation, fishing was done with the same tool as hunting. This gives an origin to a very ideal question as to whether fishing or hunting has a longer tradition? There is a marked difference between hunting and

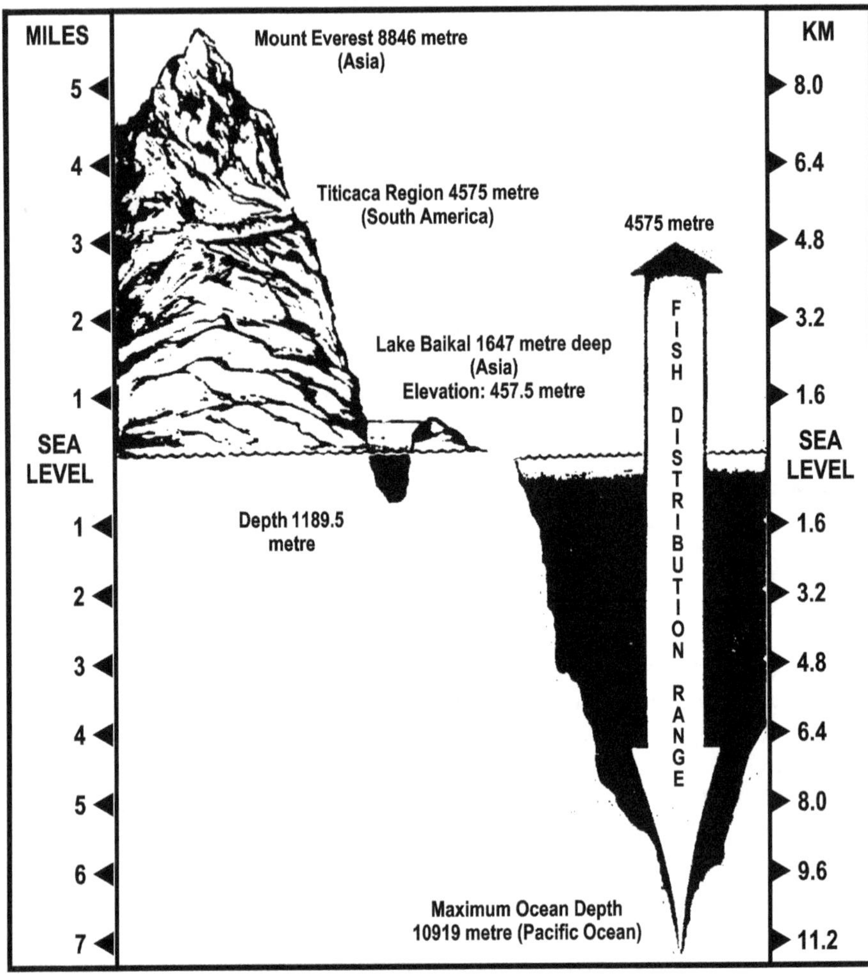

Figure 1: Extent of Fish Distribution Above and Under Sea Level (After Lagler *et al.*, 2003–modified)

fishing. The former lost its importance in most parts of the world at a fairly early stage during development and became a subsidiary element as man settled and changed to agriculture and live stock farming. Hunting also increased as a noble activity. On the other hand, fishing though has continued upto the present time but was considered virtually a profession of people of lower social standing, again with the exception of sport fishing.

Aristotle (384-327 BC) is said to be the founder of Ichthyology. History evinces that King Someswara, the son of King Vikramaditya VI who composed the book **Manosoltara** in 1127 AD was first to record the common sport fishes in India, grouping them into marine and freshwater riverine forms (*c.f.* Hora 1951a). The first modern writer on Indian fishes according to Day (1878) was Bloch whose splendid work **'Auslandiche Fische'** was published in 1785.

Fish constitute more than half of the total number of known vertebrates (39,000) recognised world over. The enlisting of exact number of fish (alive) is a highly cumbersome job. A mere scrutiny to prepare a canvas of live fish took Cohen (1970) about seven years and his estimates (based on current Ichthyological literature wherein the reflected fish number ranges from a low of 15,000 to a high of 40,000) are Agnatha–50, Chondrichthyes–515-555 and Osteichthyes–19,135-20,980. He also professed the percental distribution of living fish in various habitat (*Figure 2*).

Some past estimates of number of presently living fish species (*Table 1*) are highly varied and to these numbers are added 75-100 new species every year.

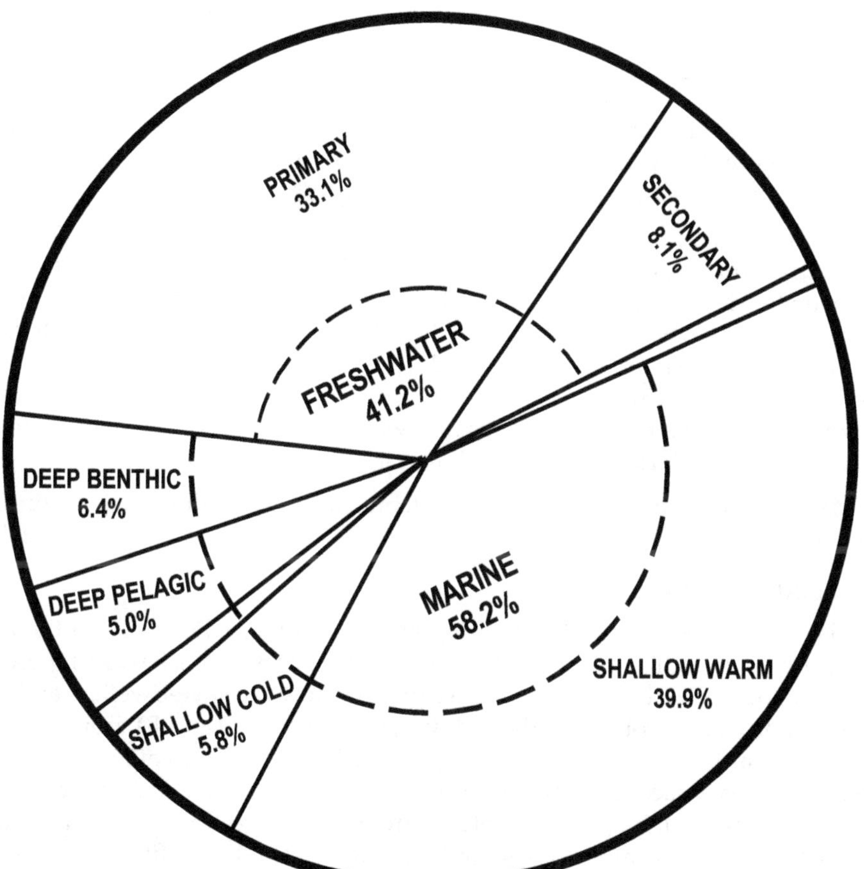

Figure 2: Percentages of Recent Fish Species Living in Various Habitats (After Cohen, 1970)

The great diversity that a look at the table suggests and Cohen's remarks as placed in his own words, "A final conclusion concerns the freshwater fishes. In view of high percentage of fishes found in freshwater and man's increasing modification of this environment throughout the world, it is vital that research be drastically increased on the basic systematics of freshwater fishes while this is still possible" speak for need of study of fish particularly the ones that live in freshwaters.

Table 1: Some Past Estimates of Number of Fish Species

Author	*Year*	*No. of fish recorded*
Schultz & Stern	1948	40,000
Myers	1958	33,000 (Teleosts)
Bailey	1960	15,000-17,000
Norman	1963	25,000
Marshall	1965	20,000
Schultz	1965	32,000

In our Indian region alone there are 2500 species of which 930 are inhabitants of freshwater and rest live in seas (Jayaram, 1999). India alone harbours 11.5 per cent of the total fish fauna so far known in the world.

Fish are of great economic value for man. Discovery of vitamin A and D during early 20th century placed them in esteemed position because of revelation that fish are generally rich in them. The food value (gastronomic, economic, culinary and nutritional) brings atleast few species in the category of "**gourmet par excellence.**" In addition to serve as an important food item, it also provides several important by-products. Fish do interest people in a number of ways. The world's smallest known vertebrate is a fish–*Pendaka pygmaea* while the largest aquatic vertebrate too is a whale shark–*Rhincodon typus* which is giant and heaviest fish.

Fishes form an important group of vertebrates for man, influencing his life in a number of ways:

As a Food of Interest

Fish has served as an important ingredient of human dietary from the time man appeared on earth, and are primarily caught for this purpose. It was supposedly one of the food of early humans (Cro-Magnan) who were cave-dwellers and fish hunters. It provides proteins, carbohydrates, vitamins (A, D and E), iron, calcium, phosphorus and other minerals. In one pound of fish the food value comes to be 300-600 calories, which is much higher in comparison to other food materials for human consumption. Its protein contents are of high digestibility, biological and growth promoting for human consumption. The digestibility of fish flesh is of the order of 96 per cent. Nutritional studies have proved that fish proteins rank in the same class as chicken protein and are superior to milk, beef proteins and egg albumen. Over and above all, fish flesh cooks easily and offers a palatable taste and flavour and is easily digestible. In fish flesh proteins constitute about 20 per cent, fats and water contents vary from 55 to 83 per cent. Besides the vitamins, minerals and enzymes, it provides energy, essential aminoacids and essential fatty acids (vitamin like function). Fish above all are rich in linolenic acid [W-3 (Omega-3) polyunsaturated fatty acids]–a factor known for its role in preventing coronary heart diseases and other cardio-vascular diseases.

For Recreation

Several species of beautifully coloured fishes are kept in the aquarium for ornamentation. From the drawing rooms of the "*Sahebs*" during the *Raj* era, ornamental fishes have invaded the homes of middle-class families giving a boost to this business and providing self-employment opportunities to millions. Presently, it is a business estimated to be of 4.5 billion US $ and the major market of ornamental fish is of freshwater origin (85 per cent) and rest (15 per cent) of marine origin. As per figures given by MPEDA (Marine Product Export Development Agency), 8 per cent of houses (86 million) in USA, 14 per cent (21 million) in Britain, 4 per cent in Belgium, 5 per cent in Germany and 20 per cent in Dutch keep ornamental fish.

Our domestic market of ornamental fish is of Rs. 60 crore, out of which Rs. 10 crore is share of 3 metropolitan cities (Mumbai, Kolkata and Madras) where 150 whole time and 1500 part time fish breeders are finding employment.

Needless to say watching fish (ornamental) is rewarding feast to the eyes and relaxation of mind and this gives pleasure to both young and old alike. Presence of aquarium becomes an attracting focal point and adds to beauty of a room where it is kept.

For Sports and Games

Fishing forms an important outdoor game for millions of people, who catch various species. Few freshwater species provide sport to anglers. Angling for recreational fishing is universally acknowledged as a thrilling sport. The money spent by anglers toward the pursuit of their hobby and by the government in maintaining streams well stocked in some of the developed countries of both west and east runs into colossal figure. Increasing number of medical authorities in USA believe that it is healthful for the people to go out for fishing. In India, angling was a favourite pursuit of Britishers and it was mainly for this reason that the trout was introduced into our country. The sport is indulged in by a section of relatively well-to-do Indians who are fond of sports and outdoor life, by the foreigners settled in India and by the tourists while on a visit to India. Trouts and all species of mahseer are the principal fishes that provide sport.

Industries

As fish forms a rich source of food, millions of people are engaged in fishing industry. Some are engaged directly in catching for markets, others are engaged in industries for refrigeration, preservation, canning, in manufacture of fish by-products.

Following are the fish products that come from industry and find use in human culture:

Sl.No.	Product	Use
1.	Fish manure	Used in fields to provide high percentage of nitrogen and phosphorus.
2.	Fish oil	Rich in vitamin A and D and used in painting, varnishing, soap, candle, leather and steel industry.
3.	Fish glue	Used as adhesive.
4.	Isinglass	Used for clearing beer and wine. Also used in making edible jelly.
5.	Silver coating of scales	Used in making artificial pearls.
6.	Fish flour	Used in making fish biscuits, fish macroni.
7.	Fish paste	Used in making sausage and ham.

Medicinal Value

The medicinal value of fish has been recognised from time immemorial. It provides a staple diet and protein supplement. The use of certain fishes in the control of malaria, in the treatment of asthma, in the prediction of impending earthquake are yet to be explored fully. The flesh of fish has less of "stroma" protein which is of low nutritive value. Fish flesh provides an excellent source of all aminoacids in concentrations needed in human diet. About 300 g of raw fish flesh suffices to supply about half of the total protein and fat requirement and 1/4th of the calories needed in a balanced daily diet for adult human beings. Fish liver oil presents vitamin A and D in proportions most suitable for human needs. Therefore, fish liver oil has been in medicinal use for a very long time as a prophylactic and curative. Its curative value, however, came to light with our growing knowledge about vitamin deficiency diseases. Fish liver oil can cure diseases caused by deficiency of vitamins A and D such as rickets, xeropthalmia, impaired vision, eye defects, abnormalities of skin etc., in the body.

Fish is replacing cattles in providing raw material for the manufacture of insulin. The pancreas of shark and whales is very rich in insulin. Fish is also rich in Omega-3-fatty acid which has a capacity to prevent blood clotting, atherosclerosis and low blood pressure conditions, in addition to lower tryglyceride levels. Cardiologists recommend that if fish is taken twice a week it will prevent heart disease. Fish can also be recommanded to convalescing patients.

An unorthodox therapy of chronic bronchial asthma is practised using fish medicine. The long queue at a private "free service" in Hyderabad once every year is a proof in itself of the efficiency of this miracle medicine to cure asthma (alive fingerlings of *Channa* sp. 2" size are used).

Electricity/Light Production

Fishes are unique in the animal kingdom in being capable of generating electric current. Fishes with powerful electric organs use them both offensively in their hunt for food and defensively against the enemy.

A number of fishes, mostly deep-sea inhabitants, possess luminiscent organs to produce light and fishes were first to generate it. That helps them to search prey in dark waters or may be defensive or used as warning signal. Ancient Romans and Egyptians practised shock electrotherapy using the power of electricity generation of some fishes.

People of certain faiths regard fish as a symbol of good omen from time immemorial. The streamlined profile of fish body has served as a basis of early attempts in the development of aircraft models.

A number of fishes possess poison glands and a stinging mechanism to serve as organs of offence or defence.

The growing realisation of importance of fish as food and of the role this fact might play in meeting ever increasing global food production in the context of immense potential of resources for exploitation has been instrumental in the initiation of development of fisheries in many countries. However, in the path of development and expansion of fisheries by a number of obstinate dogmas and customs, traditional, religious superstitions in nature. Presently, it is pity that fish has not got the importance in the diet that it rightfully deserves.

Presently, the total world catch is 70 million tonnes and about 25 million tonnes of fish annually are handled and processed in ways other than fresh, frozen, smoked, canned and fermented products which are eaten. In India, we have huge fishery potential but we are not utilising it. However, the fish production in India has registered a steady increase under successive national 5-year plans. In the period of 20 years *i.e.*, 1951-70, India's fish production increased from 0.75–1.75 million tonnes, an increase of 113 per cent. In 1985-86, the inland fish production was 36.6 per cent while 40.5 per cent inland fish production was achieved in 1990-91.

In view of highly perishable nature, fresh fish used to reach consumer only in and around production centres. During last decade, fish preservation has been developed on commercial lines.

The requirement of flesh food to meet the animal protein requirement of country are estimated as 35 gms/day, of which 17.5 gms should be met from fish. The all India per capita availability of flesh food is 13 gms of which 3.3 gms are the present contribution from fish. In other words gaps are to be made up out of increased fish supplies of 14.2 gms/day. On this basis the additional fish requirement to fill the protein gap are of the order of 1600 metric tonnes.

Thus, fish being an organism of immense food, medicinal, recreational and sport value deserves to be known both by common man and technically qualified, also for its morphological, biologial and taxonomic variables. Water with a major coverage of earth's crust and distinctive physical and chemical parameters offers a great diversity in fish. Therefore, for understanding a simple but exhaustive canvas of morpho-diversity has to be made so that these could be easily understood.

//•••••••••//

Chapter 2
Diversity in Fishes

The colourful names like angel shark, hagfish, warmouth, white fish, peacock flounder, zebra fish, flying fish, seahorse, bombay duck, climbing perch etc. given to fishes reflect their enormous diversity and as the records suggest there are about 32,000 species of fish with new ones being added on regular basis. In this chapter an effort is being made to provide a glimpse of their enormous diversity and specialities that account for their being more in number than all mammals, birds, reptiles and amphibians put together.

Names are given to fishes (like other organisms) primarily to organise our knowledge of very vast number of available species in nature, and develop an artificial classification system for them. The purpose of the system is also to show the relationship of different groups of fishes to one another.

Scientific names were and are being developed as a way of standardizing animal names, independent of common names (which may, however, tend to vary from place to place), to be accepted all over the world. Common names are highly variable, changing from language to language and within a country from region to region.

In North America, the American Fisheries Society initiated a highly successful effort to standardise the common names of fishes and published a list that is updated every ten years. These standard common names have become widely accepted by anglers, naturalists and fish biologists. Some of these names in common use are:

Lampreys and Hagfishes (Agnatha)

These are not true fishes because they lack paired fins and jaws. Their body is long, eel-like with many primitive characteristics.

Lampreys can be recognised by sucking disc on the mouth (covered with sharp "teeth"), the row of gill slits along each side of the head and large eyes (*Figure 3*). Their adults are predators on bony fishes.

Figure 3: Lamprey with Sucking Disc

Hagfishes are also worm-like with no obvious eyes (*Figure 4*). Today they are in danger of being overfished because there is an unabating demand of their skin for making leather (eel skin).

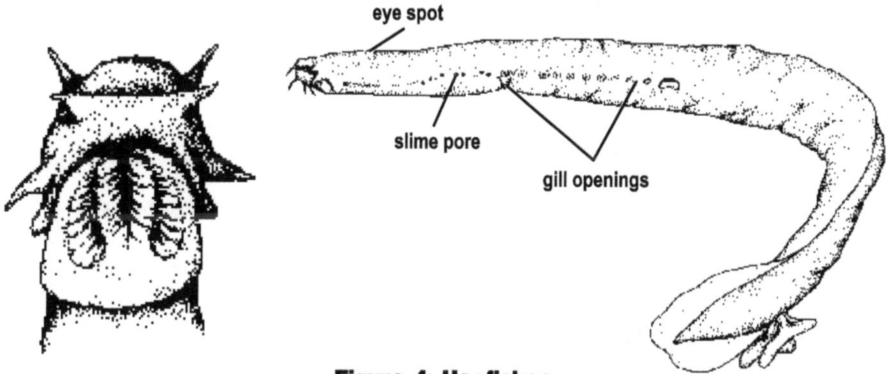

Figure 4: Hagfishes

Sharks, Rays and Skates

Elasmobranchs (class: chondrichthyes) have excellent sense of smell and their sense of hearing is very sensitive to low frequency vibrations. They possess well developed lateral line system and vision that is acute at low light levels and sensitive to colour. This group is characterised by elongated or fusiform body and heterocercel tail (*Figure 5*). Males possess claspers on tail. The skates have a depressed body, tapering snout and poorly developed pectoral fins (*Figure 6*). The rays have a flat disc-like body formed by fusion of pectoral fins. Tail is whip-like armed with pointed and serrated spine that takes the place of dorsal fin (*Figure 7*). Elasmorabranchs have 700 species and exhibit diversity in body shapes. These are the only creatures on earth who are gifted with capacity to generate electricity (*Figure 8*).

Figure 5: Shark (Spiny Dog Fish)

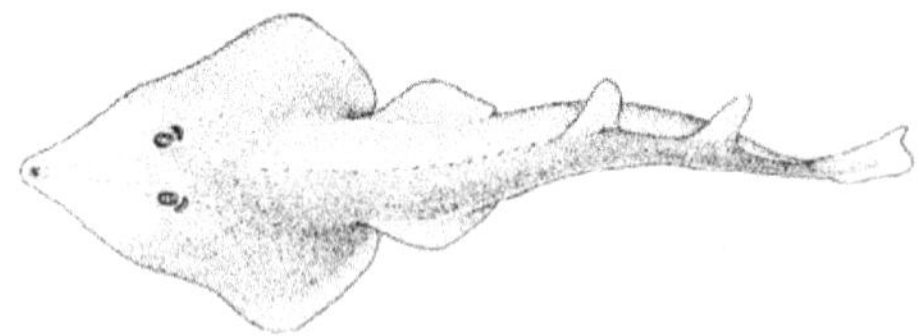

Figure 6: Skates (Guitar Fish)

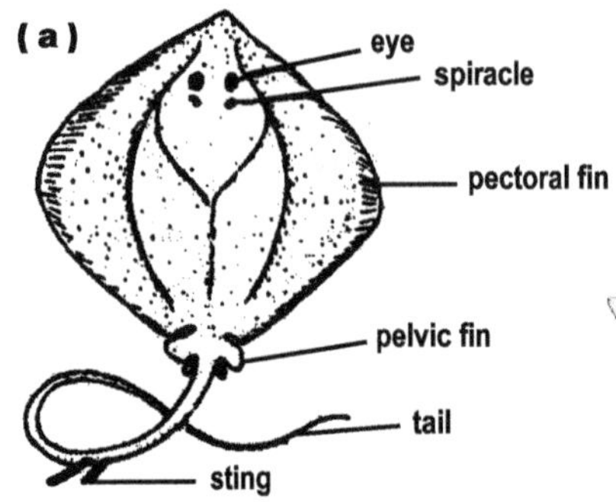

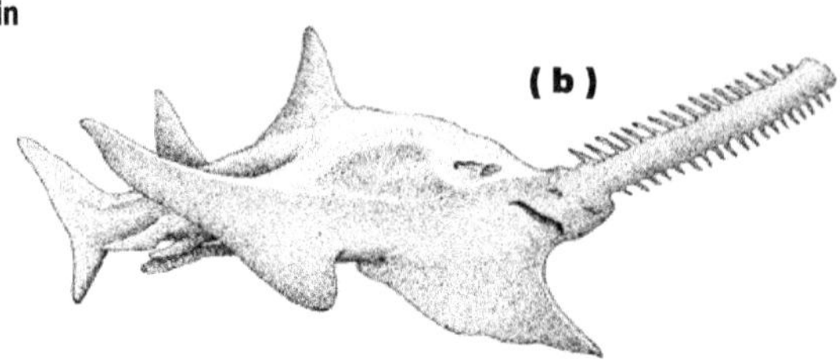

Figure 7: Rays–(a) Sting Ray, (b) Saw Fish

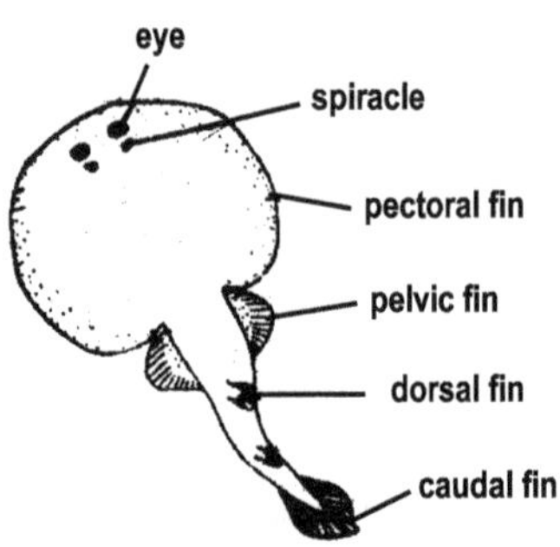

Figure 8: Torpedo (Electric ray)

Ratfishes (Holocephali)

Ratfishes or *Chimaera* species are group of 25 species of chondrichthyes with separate evolutionary history from rays and skates, and so grouped as holocephali. Their head and eyes are large but tail is long and slender (*Figure 9*). They dwell on bottom and crush their food (clams and invertebrates) with strong and flat teeth. The first dorsal fin is preceded by a spine which may be venomous and from the erect position it can be folded into a groove on the back.

Bony Fishes

Most of the world's fishes (bony) belong to osteichthyes and characteristically possess lungs (or a swim bladder derived from lungs), bone, bony scales and

Figure 9: Rat Fish (*Chimaera* sp.)

segmented fin rays. They have two evolutionary lines: (a) Tassel-fins (crossopterygii) and (b) Ray-finned fishes (actinopterygii).

Tassel-fins

During devonian times the osteolepid tassel-fins gave rise to amphibians, the coelacanth (which were once a widespread group) have only one representative *Latimeria chalumnae* (*Figure 10*) and the third group of tassel-fins, the lung fishes (dipnoi) have only seven survivors (*Figure 11*). These are characterised by limb-like fins (with supporting bones attached to pelvic and pectoral girdle). Six of the seven lobe-finned fishes are lung fishes (order: dipneusti) with one species (*Neoceratodus forsteri*) in Australia, one species (*Lepidosiren paradoxa*) in South America and four (*Protopterus* sp.) in Africa. All breathe air through true lungs which enable them to survive the drought conditions. They possess internal nostrils and cosmoid or cycloid scales.

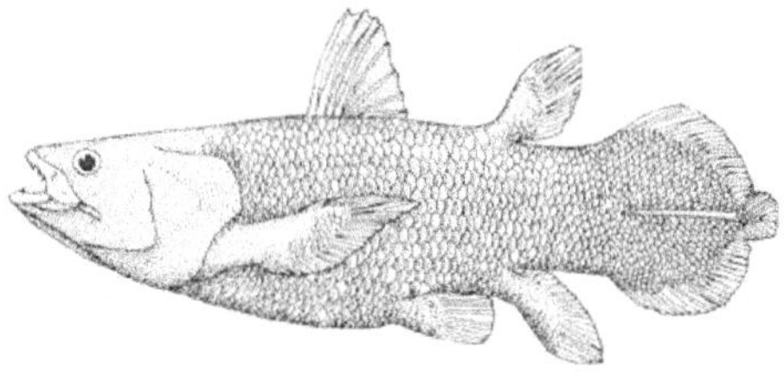

Figure 10: Coelacanth

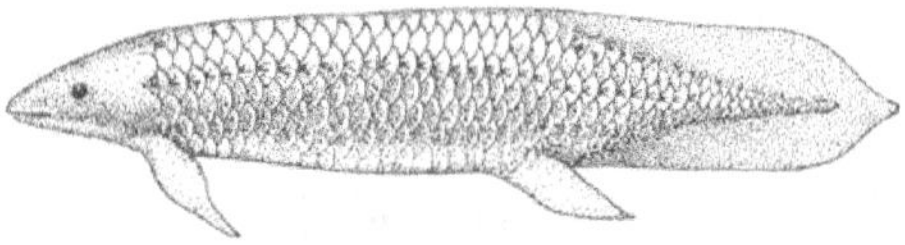

Figure 11: Lung Fish

Ray-finned Fishes

These are the modern bony fishes that outnumber all other fish groups and characteristically possess fins attached to body by fin rays (rather than lobes) and lack internal nostrils. Among other characteristics, the teleosts have a homocercal tail, thin scales, a swim bladder, highly maneuverable fins and jaws.

The diversities in some families of ray-finned fishes are as follows:

Sturgeons (Acipenseridae)

Sturgeons are primitive bony fishes that possess cartilaginous skeleton, heterocercal tail and lack scales except for a few that have been modified into ridged bony plates along the back and sides (*Figure 12*). These are among the largest fishes

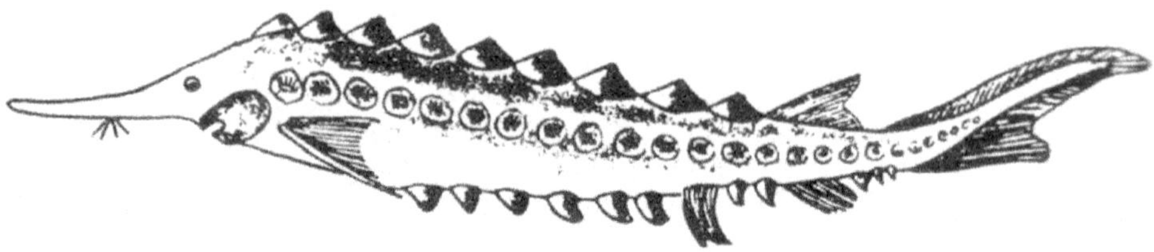

Figure 12: Sturgeon *(Acipenser* sp.*)*

occurring in freshwater and their number is declining because the eggs are prized as caviar (pickled eggs used as delicacy).

Gars (Lepisosteidae)

These belong to one of the ancient groups and are characterised by cylinderical bodies (covered with an armour of hard, diamond-shaped scales) and bony heads (tipped with long snout full of teeth for grabbing the fish they feed on) (*Figure 13*). Most of the gars reach upto 1 to 2 meters in length whereas alligator gar may reach upto 3 meters.

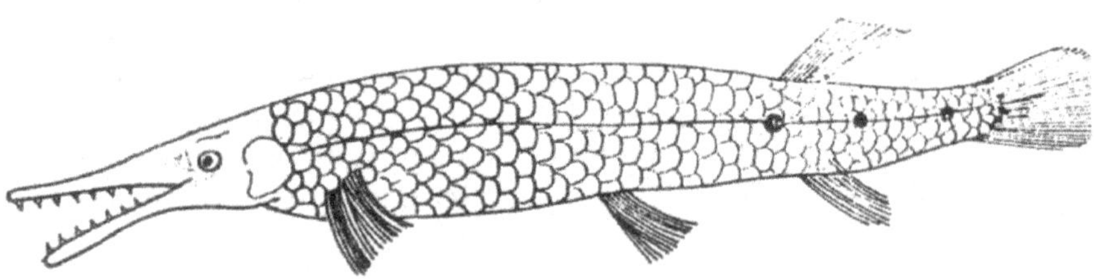

Figure 13: Gars *(Lepidosteus* sp.*)*

Herrings, Eels and Electric Eels

Herrings (clupeidae) are highly specialised for schooling and plankton feeding in open water. This family includes 300 species including shad, sardines, gizzard shad and menhaden. They can be easily recognised by their keeled (saw tooth) bellies. The large silvery scales on their body reflect light to dazzle day-time predators. The scales are also deciduous, if predator grabs a fish, the victim may escape by leaving a few of its scales in the predator's mouth. One of the characteristic features of their belly is that it tapers to a sharp keel that makes them less visible to predator attacking from below (*Figure 14*).

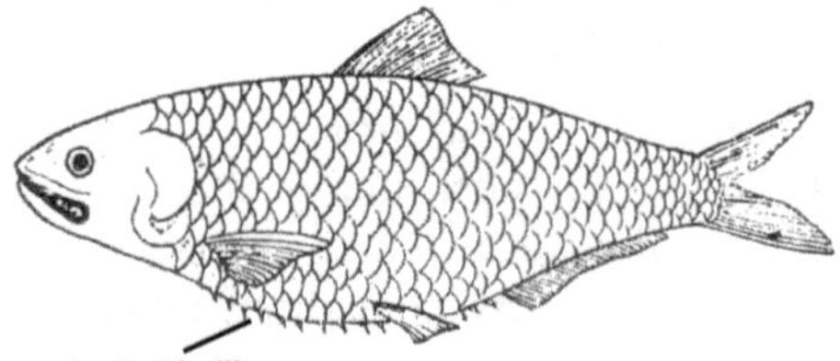

Figure 14: Herring *(Hilsa ilisha)*

Eels (order: anguilliformes) possess long tapering body with dorsal, caudal and anal fins that merge into one another and a small wedge-shaped head with sharp teeth (*Figures 15, 16*). One or more sets of paired fins are lacking and scales

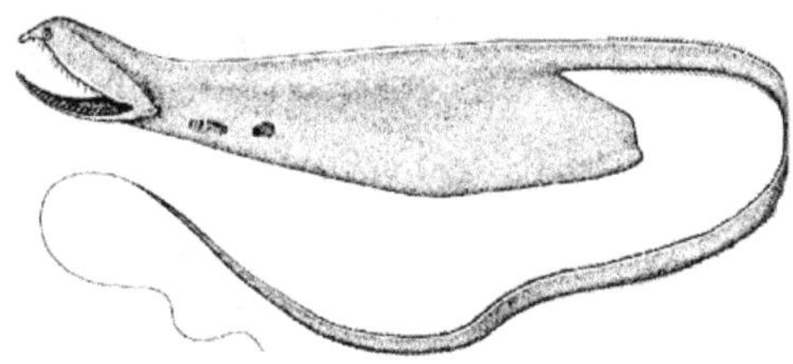

Figure 15: Moray Eel **Figure 16: Gulper Eel**

are either small or deeply embedded in skin. Eels are much more diverse than most people recognise. Moray eels are the most familiar eels because their toothy heads are frequently photographed peering out of holes in the reefs, but most studied eels are the freshwater eels (anguillidae) and members of this family are highly prized food fishes.

African Electric Fishes (Mormyridae) (*Figure 17*)

These fishes live in muddy rivers and swamps of tropical Africa where good eyesight is of little use. All mormyroid fishes possess electric organs and a special kind of skin containing sense organs, the mormyromasts, which are probably responsible for sensing electric fields.

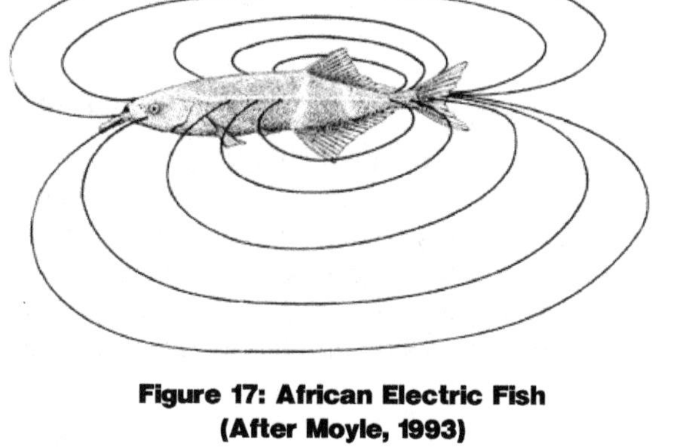

Figure 17: African Electric Fish (After Moyle, 1993)

They locate prey and communicate with one another using electric signals. These signals are produced by modified muscles in the caudal peduncle and an electrical field is set up around each fish. These signals can be used in courtship, aggressive behaviour and other intraspecific encounters. The cerebellum is large and curiously, in some species brain-to-body ratio is about the same as that of humans. The complexity of mormyrid brain is also reflected in complex behaviour patterns (*Figure 18*). Aquarists admire them for their learning abilities.

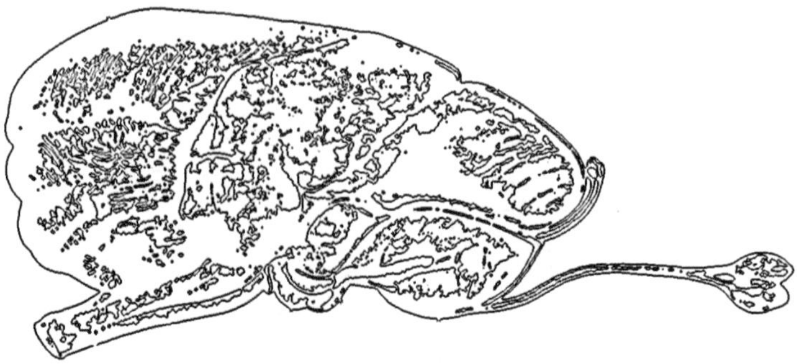

Figure 18: Mormyrid Brain

Pikes (Esocidae), Salmon (Salmonidae) and Bristlemouths (Gonostomatidae)

Pikes are important food and sport fish. They are more vulnerable to capture in spring, when they move into shallow areas of flooded vegetation to spawn. The pikes have elongated snouts and forked tails (*Figure 19*).

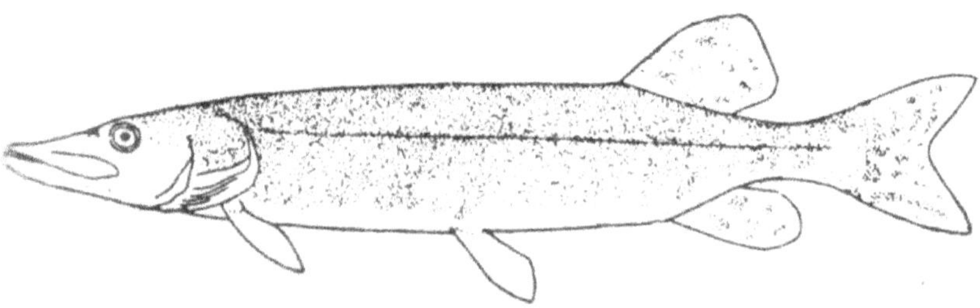

Figure 19: Pikes

Salmons, trout, whitefish and grayling (*Figure 20*) are the dominant fishes in streams and lakes (north temperate regions) and are much valuable as sport and commercial fishery. Salmonids are superbly adapted for living in cold water. Members of this family can be recognised by their streamlined bodies, forked tails, adipose fin and axillary process and by pelvic fins. Salmons and trout have fine scales, a short dorsal fin and teeth on the maxillary bone of the upper jaw.

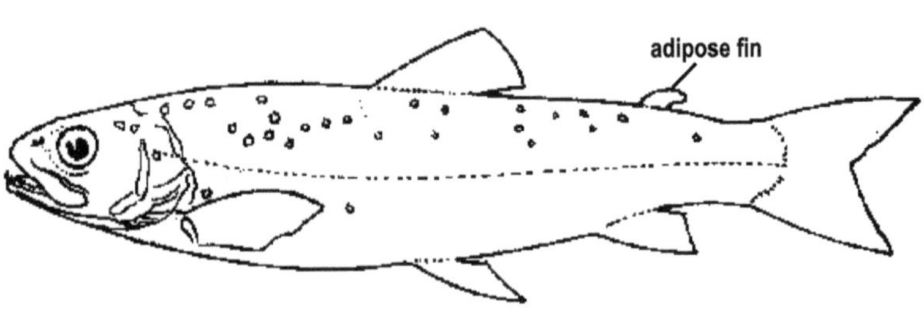

Figure 20: *Salmo gardnerii gardnerii*

White fishes have coarser scales, a short dorsal fin and no teeth on maxillary bone whereas graylings have moderately large scales, a long sail-like dorsal fin and teeth on the maxillary bone.

Bristlemouths or light fishes (*Figure 21*) are tiny (less than 5 cms) deep sea fishes with large mouths and spiky teeth and are so abundant in places that sonar beams of submarines may reflect off their swimbladder, creating impression that bottom is much closer then it really is. This phenomenon is known as deep scattering layer. They have large mouths with spiky teeth that allows them to capture a wide size range of prey, although they feed on zooplankton. Their body is elongated and somewhat rounded.

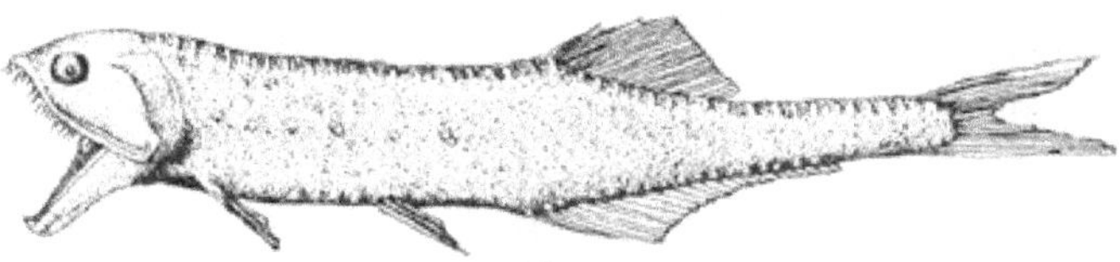

Figure 21: Bristle Mouths

Characins, Minnows and Catfishes

Characins (characidae) are a large family of freshwater tropical fishes that include some popular aquarium fishes, tetras and notorious piranhas that are lured by human blood and do not hesitate preying on man even. (*Figure* 22).

One of the reasons for the success of this group is their acute sense of hearing because of the presence of Weberian ossicles, their pharyngeal teeth (located on throat) and fear scent (chemical given by injured fish) that immediately causes its fellows to take evasive action by diving for cover or making a tight school.

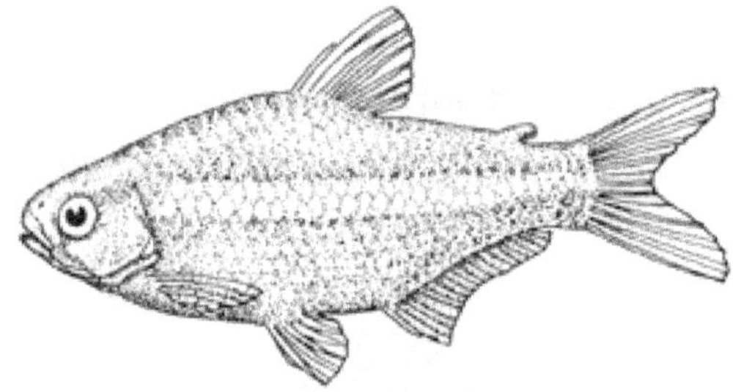

Figure 22: Mexican Tetra

Minnows and carps (cyprinidae) make up the largest single family of fishes with fusiform bodies, large eyes, visible scales and ventrally located pelvic and pectoral fins. The mouth is small with bony, flexible lips and protrusible premaxillae, which forms the upper border of the mouth and a small number of pharyngeal teeth (*Figures* 23, 24). Most species are small (less

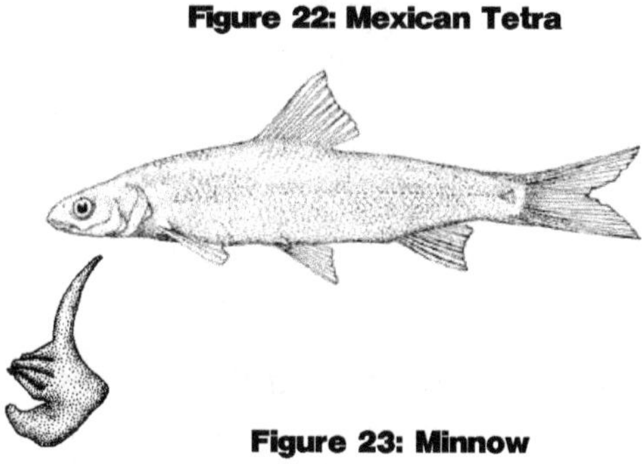

Figure 23: Minnow

than 10 cms). These fishes remain silvery most of the year but may develop bright red breeding colours in spring when the males of many species build and defend nests.

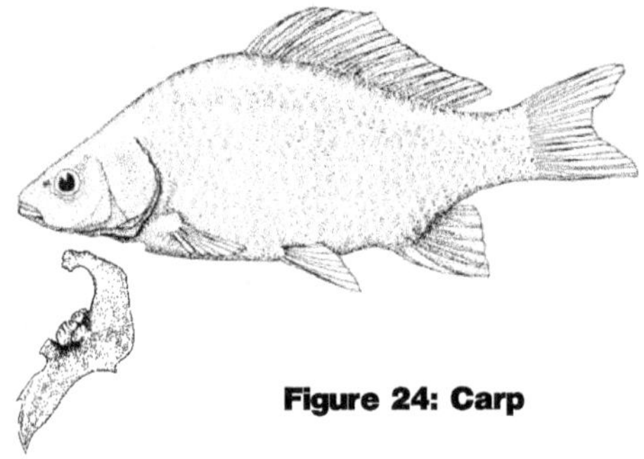

Figure 24: Carp

Suckers (Catostomidae)

These fishes are specialised for sucking and scrapping algae, and organic debris from the bottom. For this purpose they possess extended fleshy lips that point downwards, comb-like pharyngeal teeth for breaking up mats of algae and debris and long intestines for breaking down hard to digest material (*Figure 25*).

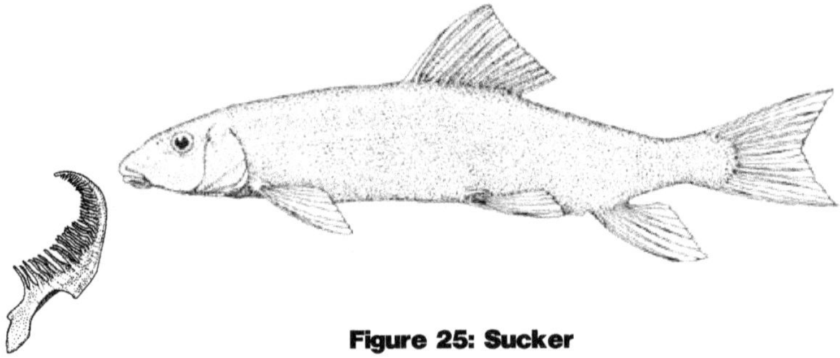

Figure 25: Sucker

North American Catfishes (Ictaluridae)

These are catfishes covered with bony plates of armour (*Figure 26*).

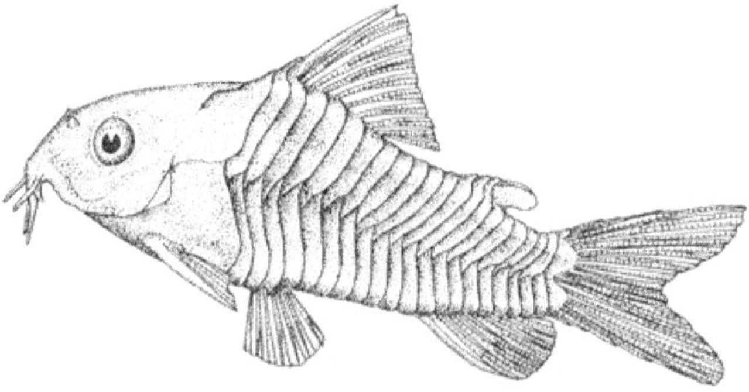

Figure 26: Armored catfish

Lantern, Cod and Angler Fishes

Lantern fishes, the members of family myctophidae represent deep sea fishes including spider fishes, paperbones, daggertooths, pearleyes and telescope fishes. They are so named because they possess row of light organs (photophores) on body and head to recognise each other in the dark. Blunt heads and large eyes are their characteristic features (*Figure 27*).

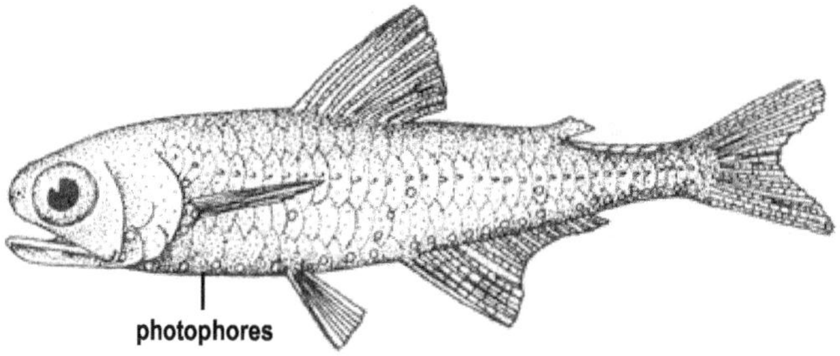

Figure 27: Lantern Fish

Cod fishes (family: gadidae) include the economically important members that are characterised by elongated bodies, three dorsal fins and a chin barbel (*Figure 28*). Their large mouths and sharp teeth indicate other fishes, crab and molluscs as diet.

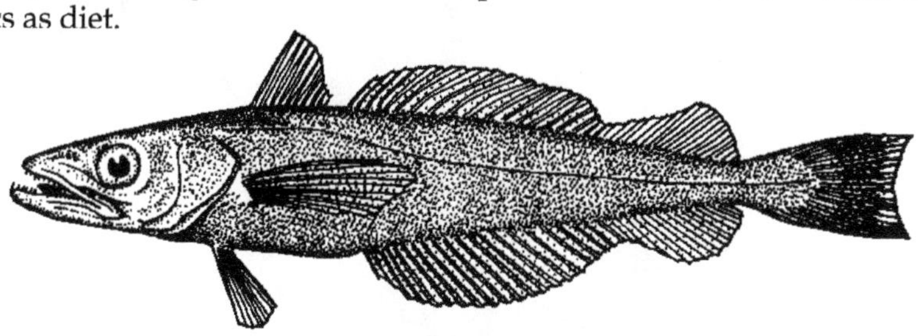

Figure 28: Cod Fish

Angler fishes (order: lophiiformes) are lumpy, flattened or globular fishes, each with a fishing pole (illicium) growing out of head and is tipped with a photophore which acts as a lure (esca) (*Figure 29*). Despite its ugly appearance (by our standards), this fish is very tasty and is sold in the market as monkfish– "the poorman's lobster".

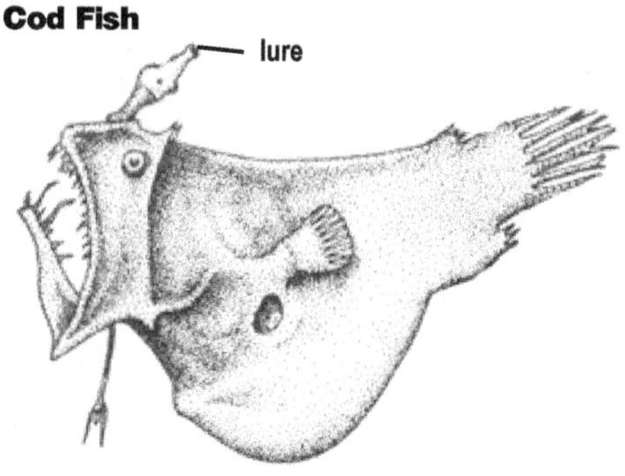

Figure 29: Angler Fish

Flying Fishes (Exocoetidae)

Unlike birds these fishes have extended wing-like pectoral fins (*Figure 30*), that are used for gliding as much as 100 metres out of water. These fishes do not use their wings to power their flight but only for gliding. The 'flying' of this fish appears to have developed as a means of escaping predators.

Killifishes (Cyprinodontidae)

These are small fishes that live in the most inhospitable environments like temporary ponds, salt marshes, desert springs, salty lagoons and even mountain

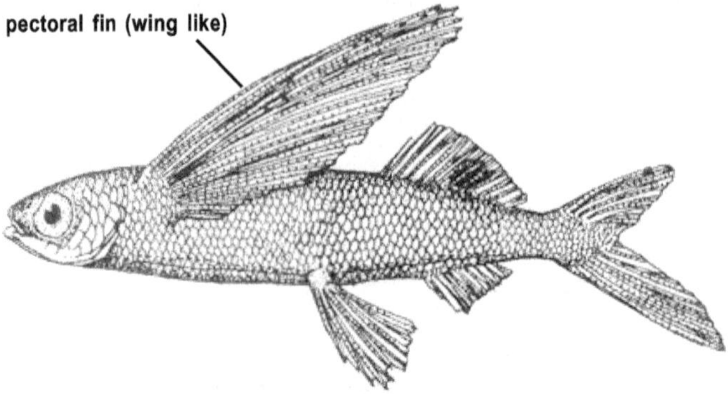

Figure 30: Flying Fish

lakes. These are brightly coloured and are favoured as aquarium fishes (*Figure 31*). Their eggs are able to resist dessication even when the water dries up. They can grow rapidly and it may take four weeks for a species to grow from egg to spawning adult.

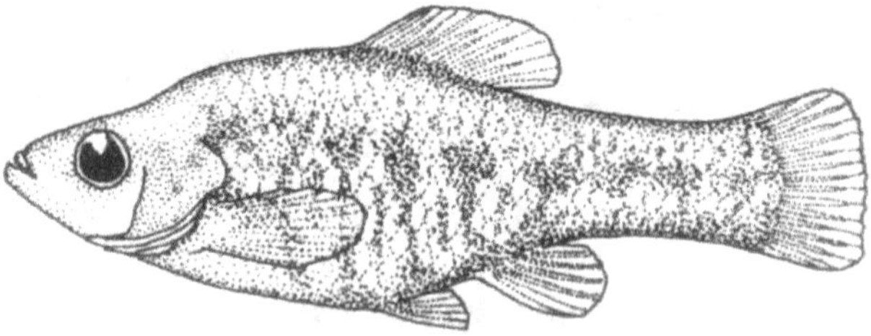

Figure 31: Killifish

In North America most of familiar killifishes are pup fishes (cyprinodon) and top minnows (*Fundulus* sp.) (*Figure* 32). These give birth to young ones (poeciliidae) and hence are called live-bearers. This family includes hardy aquarium fishes as guppies, sworld tails, mollies and mosquito fish.

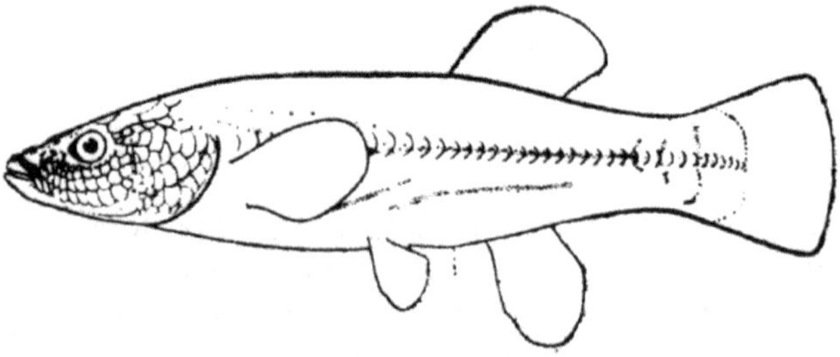

Figure 32: Pup Fish

Sticklebacks

Sticklebacks (family: gasterosteidae) are the most recognizably fish-like members of otherwise peculiar looking order gasterosteiformes (*Figure* 33). These fishes are characterised by extremely elongated body (tubesnouts, pipe fishes, seahorse, trumpet fishes). Sticklebacks are small freshwater fishes with sharp spines on the back and a row of bony plates on each side in lieu of scales. Seahorse (family: syngnathidae) (*Figures 34, 35*) are pipe fishes whose bodies have assumed a permanent S-curve, so they can swim upright, moving by sculling with their dorsal fin. Parental care is well established in them, where male possesses a brood pouch and female transfers the eggs into this pouch which remain there till hatching (*Figures 34, 36*).

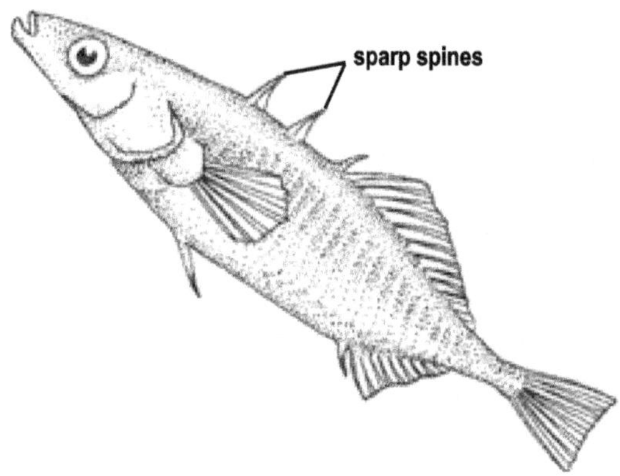

Figure 33: Sticklebacks

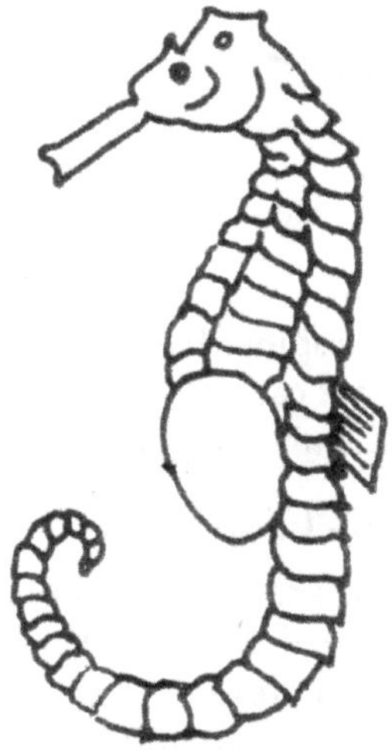

Figure 34: Hippocampus with Brood Pouch

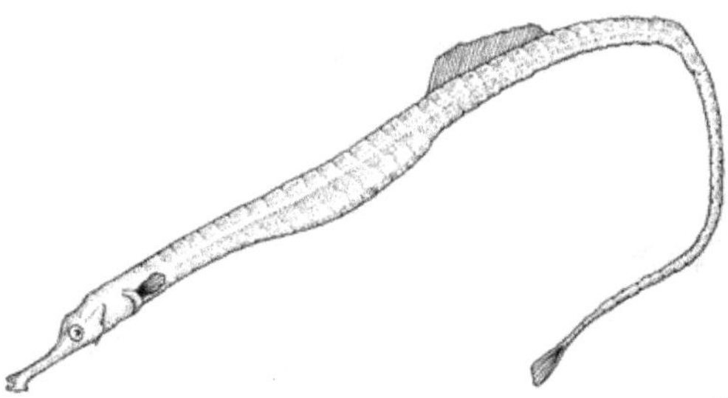

Figure 35: Pipe Fish (Female)

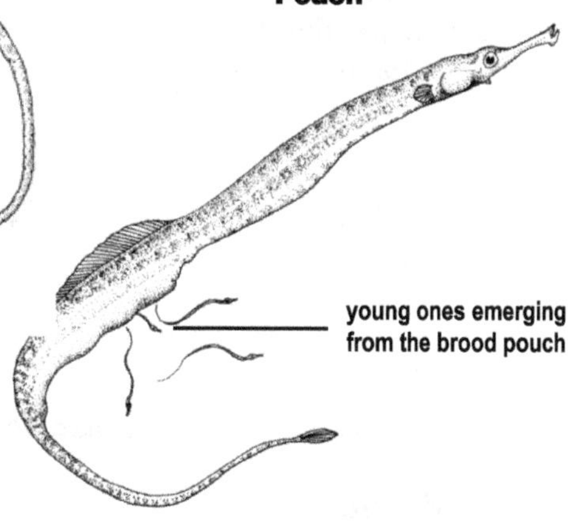

Figure 36: Pipe Fish (Males Carry the Embryos and Young in a Special Pouch)

Rock and Scorpion Fishes (Scorpaenidae)

It includes sea robins, poachers, snail fishes, lump suckers, all of which have a close attachment to the bottom. They possess large heads and mouths, large round pectoral fins, rounded caudal fin and extremely spiny dorsal and anal fins (*Figures 37, 38*).

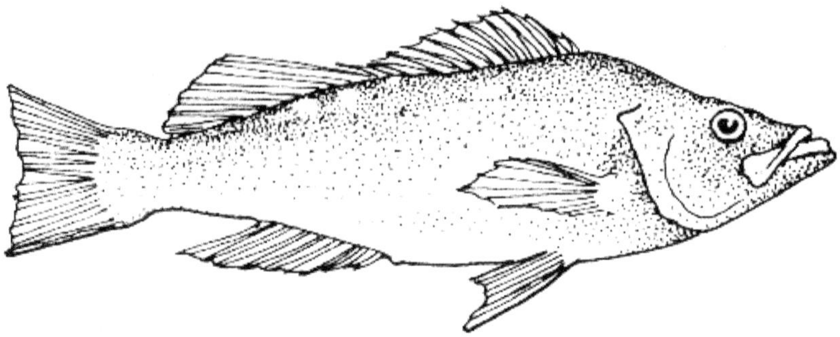

Figure 37: Yellowtail Rockfish

These spines have venom glands that can cause wound or even death of a person.

Figure 38: Scorpion Fish

Sculpins (Cottidae)

These are small, smooth, large-headed fishes that are found in environments rich in food but very turbulent. So they are adapted for living underneath rocks or clinging to bottom while the currents sweep over them (*Figure 39*).

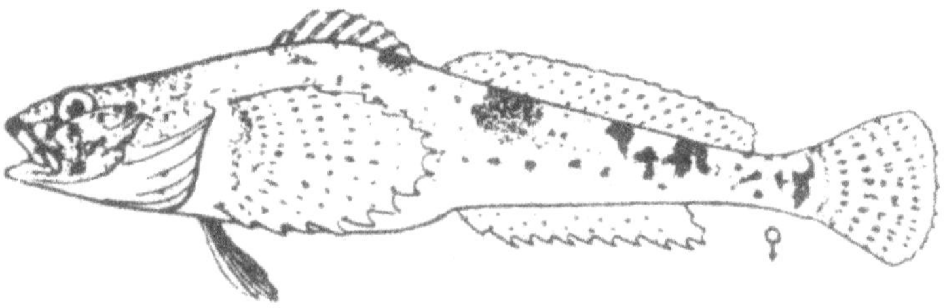

Figure 39: Sculpins

Ruling Perches

Sea basses (serranidae) and temperate basses (percichthyidae) are families of order perciformes. The basses like other perciform families possess spines on all fins (except the caudal) with two parts of dorsal fin and the lead half is made up of spines *(Figure 40)*. Pectoral fins are inserted vertically on the side of the body with pelvic fins located below them and they possess ctenoid scales, a swim bladder that serves as an organ of buoyancy, a number of serranids start out as female when small, but change into males as they get larger. The temperate basses are, however, incapable of sex change.

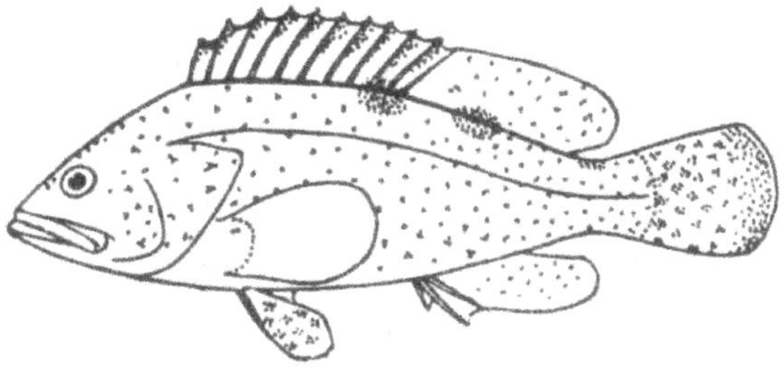

Figure 40: Ruling Perch

Sun Fishes and Black Basses (Centrarchidae)

It includes important sport and aquarium fishes such as large and small mouth bass, blue gill, redear sunfish, dwarf sunfishes, black basses and white crappies. Centrarchids are related to sea basses and temperate basses and are separated by skeletal differences (differentiated by ichthyologists only) (*Figure 41*).

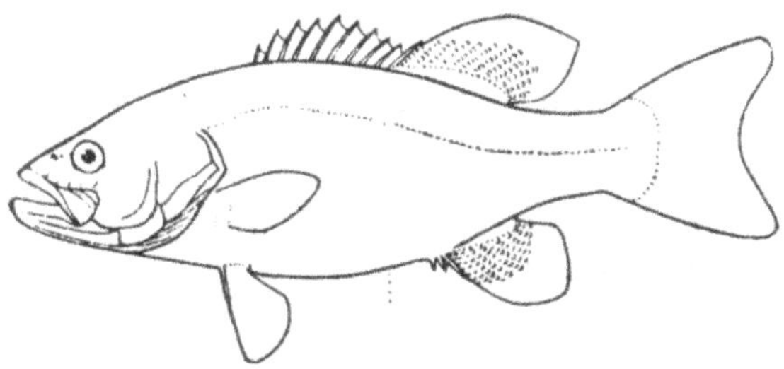

Figure 41: Basses

Perches and Darters (Percidae)

It includes highly priced game and food fishes. They are characterised by large eyes lined with light gathering layer (tapetum lucidum) that makes it possible for them to hunt other fish under dim light conditions (*Figure 42*).

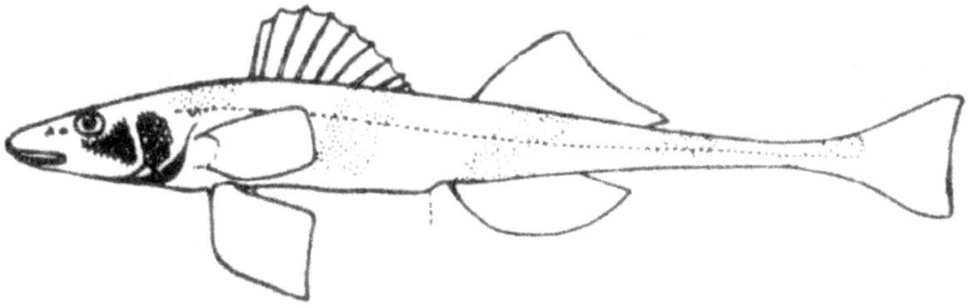

Figure 42: Perches

Drums (Sciaenidae) or Croakers

These are heavily bodied clam crunchers with strong pharyngeal teeth (*Figure 43*). The name 'drum' comes from loud sound that they make by vibrating their swimbladders, as part of their courtship ritual. They have exceptionally large ear stones (otoliths) to receive sounds.

Surf Perches (Embiotocidae)

In appearance, they seem to be more or less standard deep-bodied perciform fish, much like sunfishes. They give birth to young ones (viviparous) (*Figures 44, 45*).

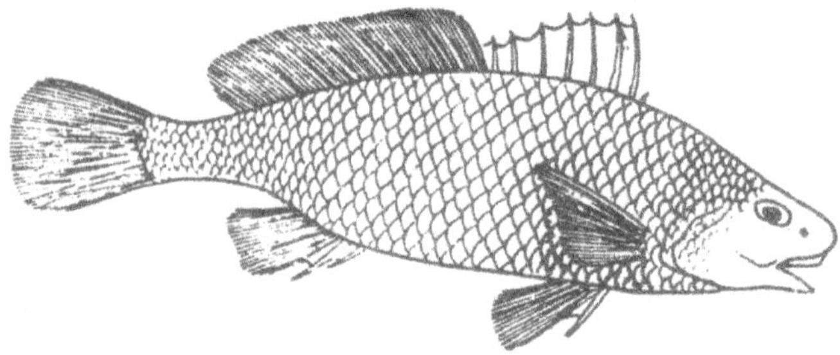

Figure 43: Drums

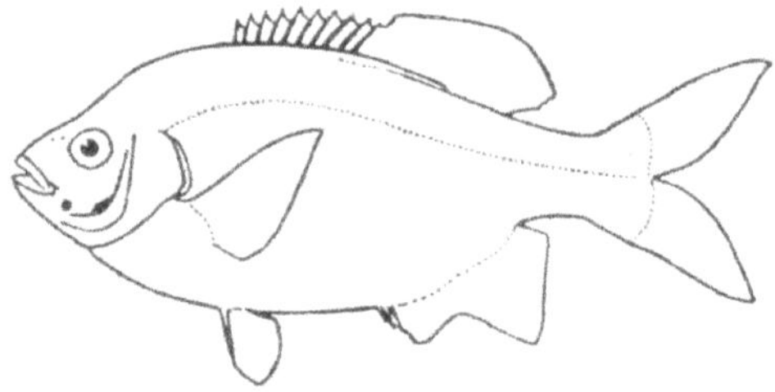

Figure 44: Surf Perch

Figure 45: Black Sea Perch

Cichlids (Cichlidae)

They possess unique and complex structure of pharyngeal teeth and associated muscles, combined with highly specialised jaw teeth. They have two sets of specialised jaws, one for obtaining the food and other for processing it. Parental care is well exhibited by them. In tropical lakes, they are the dominant fishes (*Figure 46*).

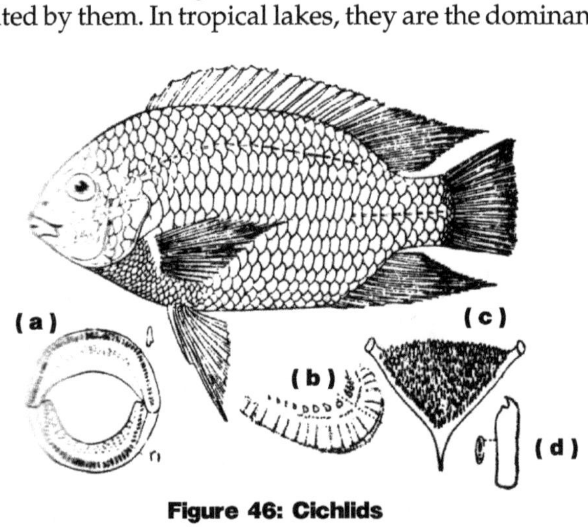

Figure 46: Cichlids

(a) Arrangement of Teeth on the Jaws; (b) Gill Arch; (c) Pharyngeal Teeth; (d) Form of Teeth

Wrasses (Labridae) and Parrot Fishes (Scaridae)

These two are closely related families of reef fishes. Wrasses are small, cylinderical fishes with small mouth containing tiny "buck" teeth (*Figure 47*). Wrasses and parrot fishes have most complex mating system with several types of males and individuals changing sex on short notice.

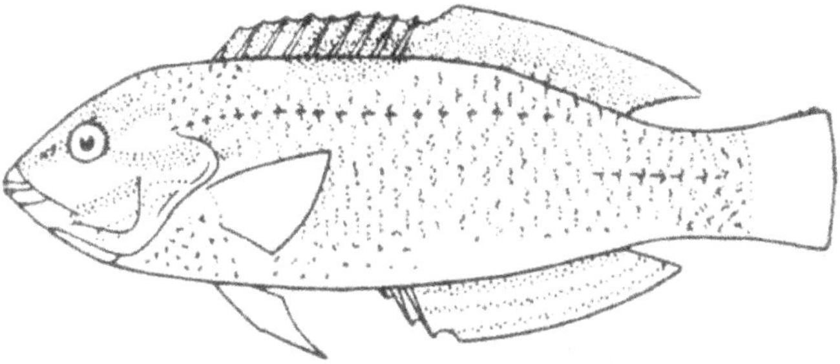

Figure 47: Wrasses

Gobies (Gobiidae)

These are second only to cyprinids (with 800 plus) species in a family. These can be recognised on the bases of, rather, elongate body with blunt heads, bulging eyes, rounded tails and most important, pelvic fins united to form a sucker. Many gobies are small and one Phillipine species (*Pendaka pygmaea*) enjoys the distinction of being world's smallest vertebrate. Maturity in these fish is acquired when these are just 6-12 mm long (*Figure 48*).

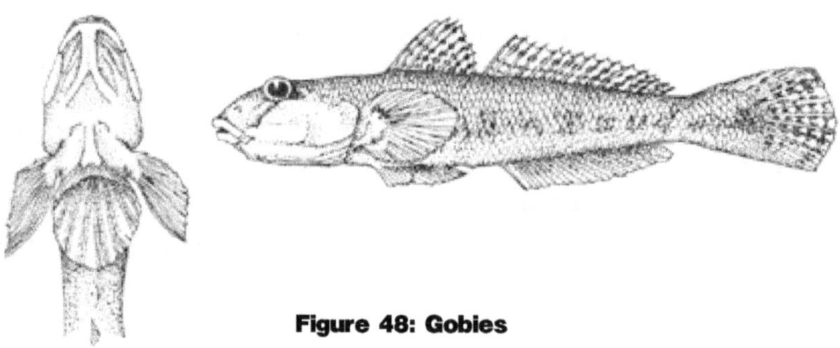

Figure 48: Gobies

Tunas and Mackerels (Scombridae)

They are the most valuable commercial fish species. They are extraordinarily adapted for a life of continuous swimming. Their body is beautifully streamlined, with smooth skin having tiny embedded scales and an iridescent hue. Dorsal fin possesses stout spines for defence, but they can be lowered into a groove along the back for rapid swimming. If a tuna stops swimming, it suffocates (*Figure 49*).

Flat Fishes and Fugu

Flounders and soles are the only fishes (pleuronectiformes) that spend their life lying on one side. The mouth in them acquires a twist to improve bottom feeding

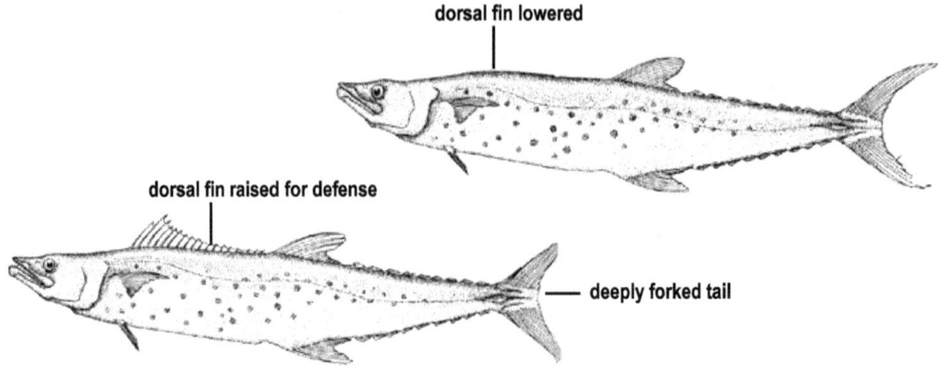

Figure 49: Mackerel

efficiency. The side on the bottom is white and eyeless, whereas other is pigmented with both eyes (*Figure 50*). The eyes may be present on left side, right side or on either side. There are about 500 species of flat fishes and some of them are important commercially.

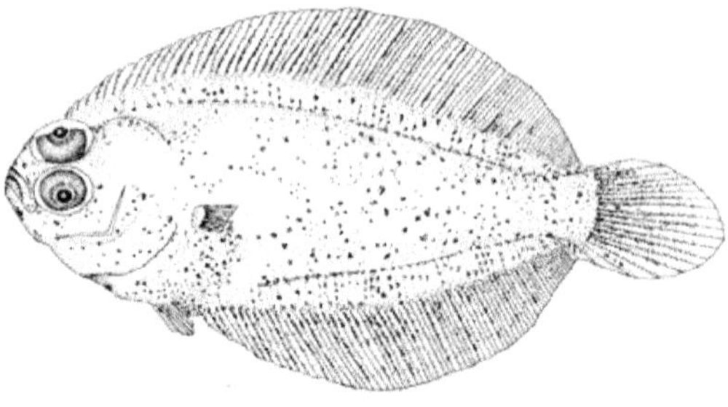

Figure 50: Flat Fish

Puffers, Box Fishes and Porcupine Fishes (Order: Tetradontiformes)

Slow swimming and heavily protected fishes with strong beaks are included in this group. To make themselves less palatable to predators, most of the members of this group can puffy themselves up with large quantities of water (*Figure 51*), which not only makes them physically large but causes the spine in their skin to bristle. Though some species are poisonous yet they are the most prized food fishes in Japan, as fugu. Fugu is non-poisonous if properly cooked. So only licensed cooks are allowed to do so. Because cooks make mistakes on occasions, the flavour of this fish is enhanced by the danger involved in eating it.

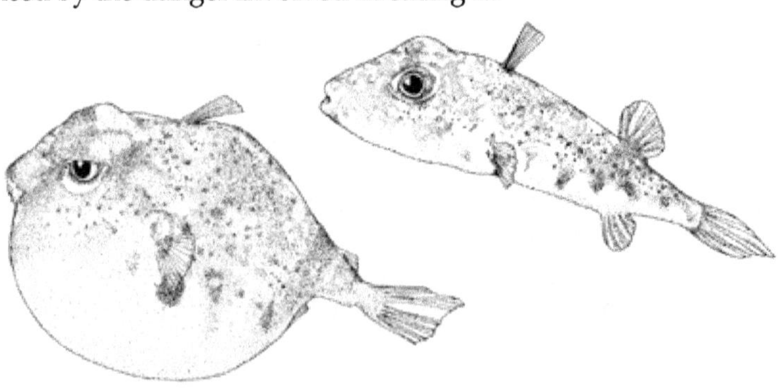

Figure 51: Puffers

//••••••••••//

Chapter 3
Collection Methods

For a scientific or taxonomic study, the best way to collect fish is to catch them alive through any of the various fishing devices in use. Area of collection should be selected where water is in abundance and help of local fisherman can be taken for collection. Cut, damaged, mutilated or spoiled specimens should not be considered and hence discarded because only fresh specimens last long after preservation.

Fishing was done in the days of yore by hand at places where water course was dammed resulting in high fish concentration to permit catching. Fishing gears were developed later which have subsequently undergone evolution in different parts of the world giving rise to various methods in use at present times. The method in use at a particular place thus depends upon the type of fish to be caught, nature of bed and available capital.

Gears and crafts are the real means of fish capture in fishing technology. Gears are the instruments used for fish capture and crafts provide platform for fishing operations, carrying the crew and fishing gear. Separate type of crafts and gears are operated in inland waters and sea. Sometimes gears can be operated without the aid of a craft.

Over the years, traditional fishing gears have been upgraded and new and more efficient fishing systems have been introduced. Some significant developments that have taken place to improve upon gears and crafts and their efficiency include: (*i*) introduction of synthetic gear; (*ii*) acoustic fish detection; (*iii*) satellite based remote sensing techniques; and (*iv*) mechanisation of propulsion.

A 'fishing unit' comprises of fishing gear along with vessel, auxiliary equipment and men. Its size can be determined by the distance of fishing ground from the shores, handling and disposal of catch as well as geographical factors. Amount of catch by a unit depends upon its efficiency and productivity of fishing grounds. Most of the fishing gears have to breakoff operations for rest and repair. The entire period between launching of fishing gear and launching it again after a gap is called as "fishing cycle" (the number of fishing cycles per day depends upon density of fish, type of fishing methods used, geographical conditions of fishing grounds and fishermen).

There are many different methods of fish collection and some of these require highly sophisticated equipments and capture skill. The most practical scientific collection method is netting. The gears and crafts used for this purpose in various parts of our country are indigenous, non-mechanised and locally built and their type of operation depends on the nature of waterbody and type and age of fish species.

Several systems of classification of fishing gears has been developed on the principles of capture, design and technical features, and operational methods. Fishing gears, whether primitive or sophisticated, revolve around five basic capture process *viz.*, gilling and tangling (*e.g.*, gill nets and trammel nets), trapping (*e.g.*, traps), filtering (*e.g.*, trawls, seines and other net fishing systems), hooking and spearing (*e.g.*, hook and line harpoons) and pumping (*e.g.*, fish pumps). Among the variety of harvesting systems available around the world, the most significant are purse seines and trawls, followed by lines, gill nets and entangling nets and traps in the commercial fishing.

Fishing gear can be described as active or passive. Active fishing systems are generally energy intensive and more productive than passive gears. Depending upon the sector in which they are used, there are small scale or artisanal fishing gears covering traditional low energy system of fish capture and large scale industrial mechanised fishing system (*e.g.*, purse seines, automated long lining etc.).

Based on waterbody in which they are used, the fishing gears are inland fishing gears (including riverine, estuarine, reservoir gears) and marine fishing gears.

Based on the area of operation, these gears may be coastal, offshore and deep sea fishing gears. Depending upon fishing position in water column these gears may be pelagic, midwater and demersal fishing gears.

Brandt (1984) grouped the fishing gear and methods into thirteen categories *viz.*, fishing without gear, wounding gear, stupefying method, line fishing, fish traps, traps for jumping fish, bag nets with fixed mouth, dragged gear, seine nets, surrounding nets, dip or lift nets, falling nets, gill nets and tangle nets. In International Standard Statistical System of Classification adopted by FAO for fishery statistics, fishing gears are grouped into fourteen categories *viz.*, surrounding nets, seine nets, trawls, dredges, lift nets, falling gear, gill nets and entangling nets, traps, hooks and lines, grappling and wounding gear, harvesting machines, miscellaneous gear, recreational gear and gear not known or not specified. Major categories of fishing gears are as below:

Active Fishing Gears

Surrounding Nets

These nets are roughly rectangular walls of netting rigged with floats and sinkers, which after the detection of the presence of fish are cast to encircle the fish school. Surrounding nets are generally operated in the surface layers.

Purse Seines

These are the predominant type of surrounding nets used for catching pelagic fishes by encircling a school of fish (*Figure* 52). It is operated from the vessel and landing also takes place on the vessel. In the hung condition, the line of the net-wall remains on the surface and lead line remains at certain depth without touching the bottom. A purse line of strong cables is also there in addition to lead line that prevents fish escape during hauling. The net essentially consists of a single wall without bag. Its central part acts as the bunt which has the thickest twines. The net is operated from vessel when one end of net is held on a small boat and then the net is laid out

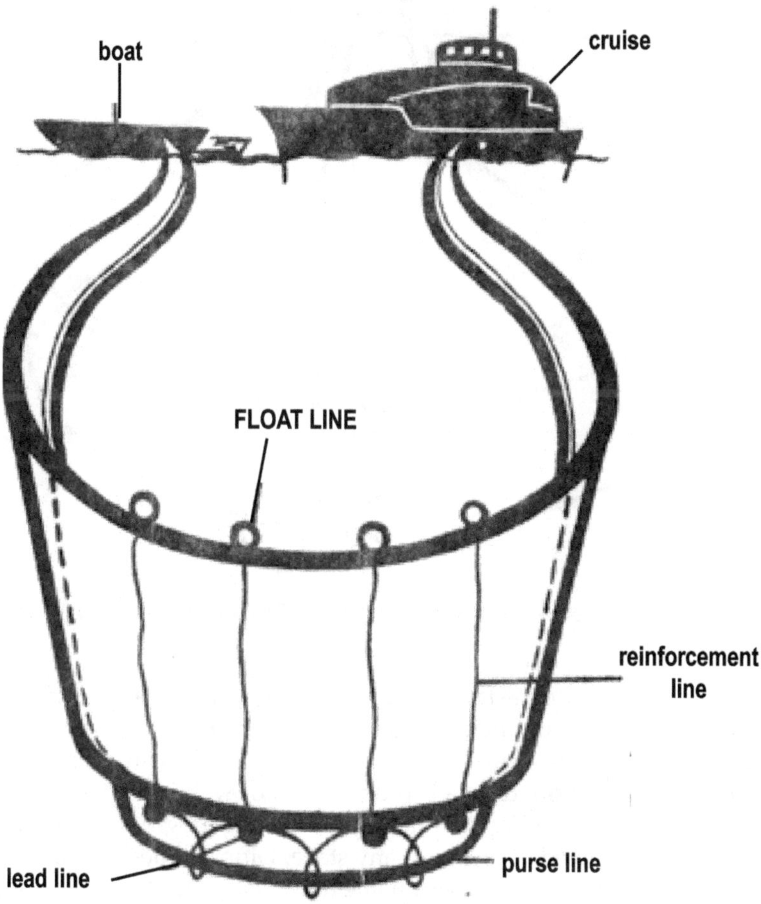

Figure 52: Purse Seines

by vessel cruising at a higher speed and after making circle comes back to the small boat (skiff). The hanging net now makes a circular wall around the shoal of fish. Now the purse line is hauled on board the vessel. As this progresses, the net takes the shape of a purse. Finally the entire net is hauled aboard along with the catch. Purse seines are the largest nets which contribute to major share of the landnigs. Some of them have a total length of 600 meters, height of hanging net of 100 meters and a weight of 3000 kg. Purse seines can be classified on the basis of number of vessels used in operation (one boat or two boat purse seines), or on the basis of fish species to be caught (anchovy/sardine/mackerel/cod/tuna purse seine). These nets may also be classified on the basis of scales of operation as small, medium or large purse seines. Surrounding nets without purse line like lampara net are operated in small scale sector (*Figure 53*).

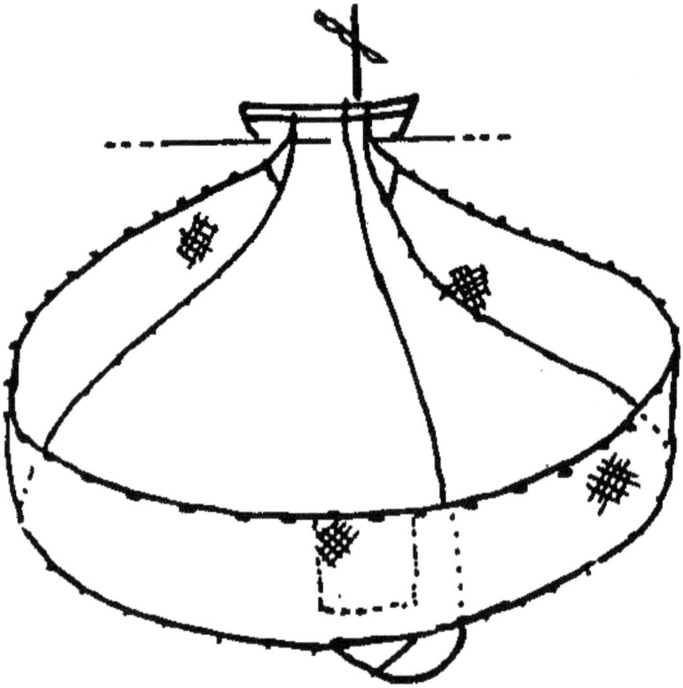

Figure 53: Surrounding Net Without Purse Seine

Seine Nets

Seines are some of the largest nets with long wall of netting with or without a bag, supported by floats and sinkers. The net is operated by ropes attached to the end of wings which are used for hauling and herding the fish. This net works on the principle of first surrounding a certain section of water mass within the net wall and then filtering that section of water and as hauling progresses all fishes are brailed out. Fish is not allowed to escape at any stage. Different designs of seines are used in lakes, rivers and marine waters. Some nets may have 1800 meters long net wall with a width of 12 meters and a total weight of 5500 kg. They are usually operated in coastal or shallow water where bottom and/or surface act as natural barrier.

Seines can be operated from boat called boat seines or from shore called shore or beach seine.

Shore Seine

Large sized nets made up of 100-500 rectangular pieces joined end to end. Each piece may be 3-6 m long. Side pieces are larger (3-5 cm mesh size) and middle pieces are smaller (3-5 cm mesh size). Floats and sinkers are provided to the upper rope and lower ropes, respectively. Long ropes are attached to both the ends for hauling. One end of net is held on shore by 20-25 men. The net is carried on the boat in a semi-circle to enclose large area and then brought back on the shore, where it is held by group of 20-25 men. The two ends are pulled and net is dragged to shore. The large net of Karwar coats is known as 'Rempani net' (*Figure 54*).

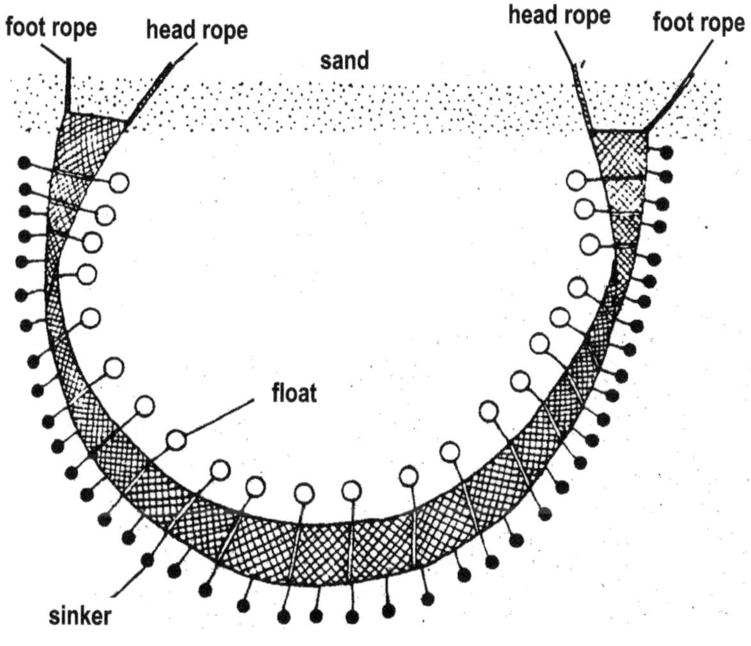

Figure 54: A Rampani Net

Boat Seine

Boat seine are conical bags with wings. The bag is 8-10 m long and is made up of circles of netting of various mesh sizes. The wings are 30-50 m long. The mouth is kept open with the help of floats and sinkers. Two boats carrying the net to the desired fishing area and on reaching there, they move away from each other in semicircular manner paying out the net. The towing ropes attached to the ends of the wing are tied to the boats. After moving around the shoal for sometime, two boats come closer and net is hauled. Similar nets of smaller sizes are used at different places and even named differently. This boat is operated in sea by catamarans. The fishes are trapped in the bag and coir ropes are used for the purpose of towing (*Figure 55*).

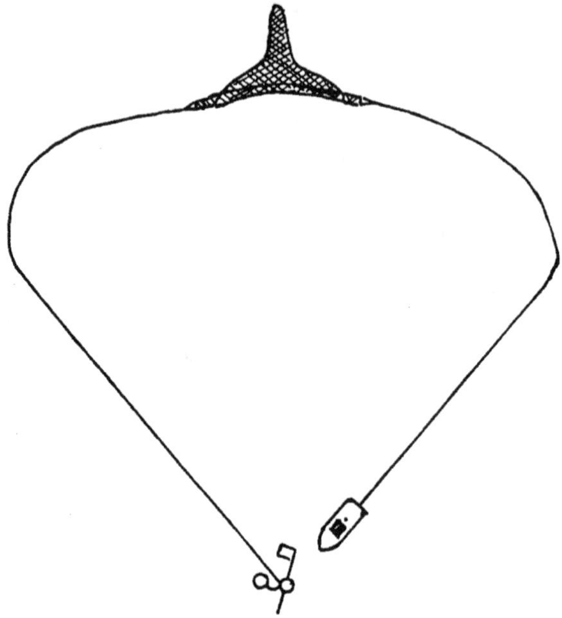

Figure 55: Boat Seine

Beach Seine (Haul Seine)

It consists of two walls of net webbing of strong twines which are known as wings. The wings are joined to the central bag of net-webbing. Smaller seines (catch capacity less than 100 kg/haul) may be without bag and called drag seine. The wings may be equal or unequal in size tapering at the ends. At the ends, they are connected to warps either directly or by means of brail of stout pole. At regular intervals, spreaders are present which run from float line to lead line. The float line carries suitable floats and the lead line has sinkers (*Figure 56*). The net is laid in such a way that its one wing remains at beach. The running wing is shot, at right angles to the shore. This end is again brought back to the beach. As a result, large arc is formed and it encircles a particular section of water. The float and lead line do not allow fish to escape. Hauling is carried out by pulling the two warps and assembling

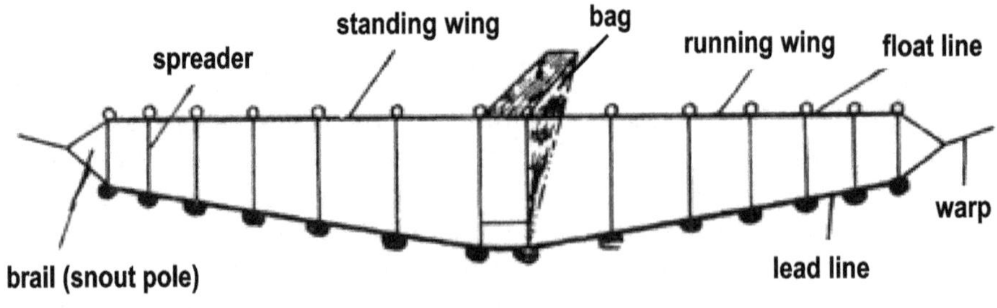

Figure 56: Beach Seine: The Net

the wings. This is done by using man power or machine power. The landing is collected from bag finally.

Danish Seine (Drag Seine)

These are like beach seines with an almost same principle of operation and are used in deep water. The essential difference lies in the fact that float line remains under water and never reaches surface. The wings are smaller and mostly replaced by warps.

Trawls

These are one of the commonly used large nets throughout the world (*Figure 57*). These nets help in catching the fish by filtering water as the net is towed. These are conical bag nets with two wings and a codend where catch is concentrated. The nets are operated by towing from one or two boats. The upper margin of mouth is

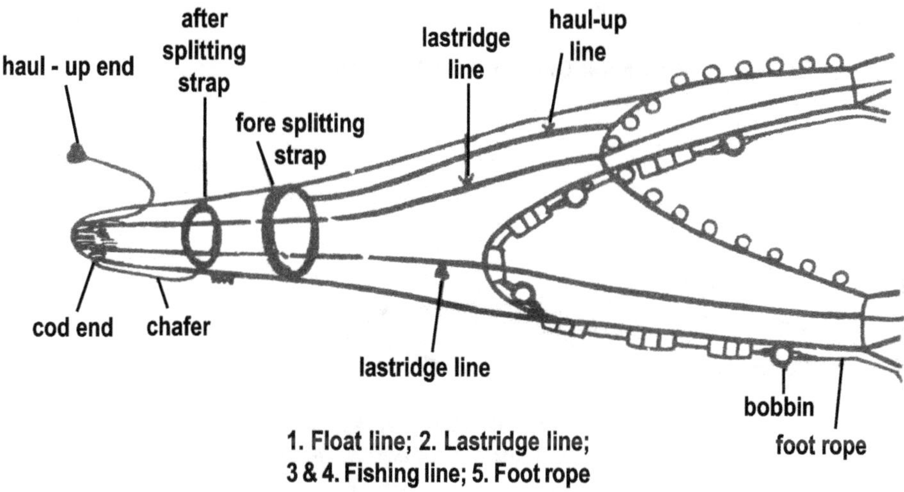

Figure 57: Trawl: The Gear

lined by float line and the lower margin by foot rope. The float line is important in keeping the mouth open. Hydrodynamic floats are used nowadays. Foot rope is rigged with a number of bobbins (wooden or metallic) and is connected to fishing lines. This arrangement prevents wear and tear of the net during dragging of the trawl net. At intervals the bag carries splitting straps to prevent its collapsing. From the side angles of mouth extends net webbing in the form of wings. Beyond the wings the main line extend as bare ends (leg of lines). The length of net is expressed by total length of float line and length of net webbing is one to two and half times that of float line. To assist in towing and dragging special rigging is carried between the warps and legs of the line of the net. Important among these include otter boards, sweeplines and spreaders. On the position of their operation in water column, trawls may be classified as bottom trawl, mid water trawl or pelagic trawl. Based on the number of boats there may be one boat (*Figures 58, 59*), two boat trawl or pair trawl or bull trawl (*Figures 60, 61*).

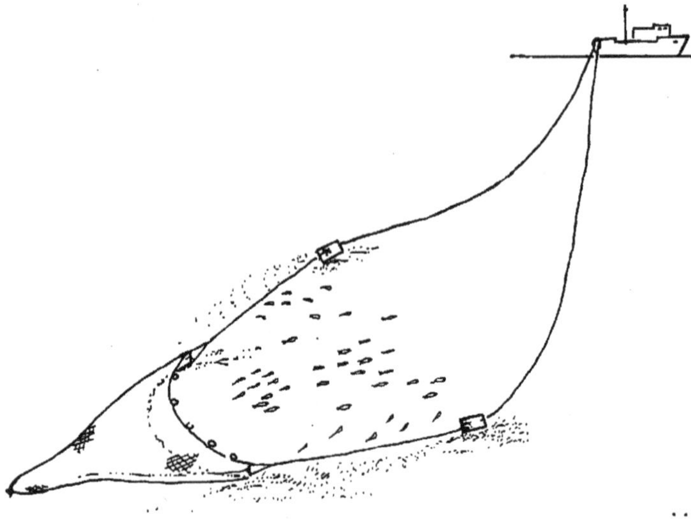

Figure 58: One Boat Bottom Otter Trawl

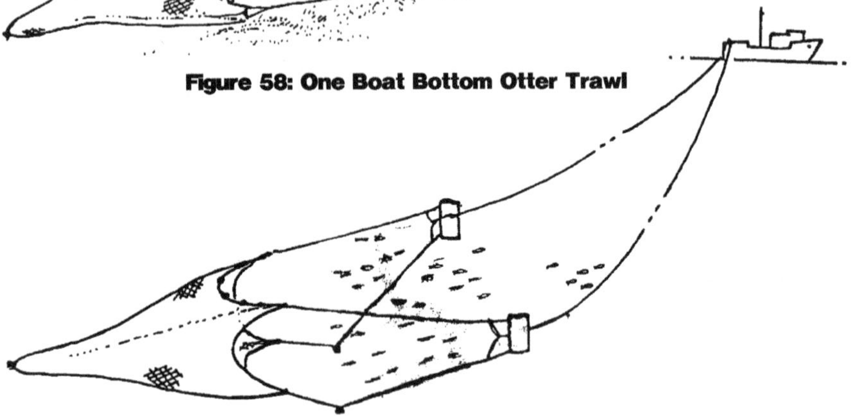

Figure 59: One Boat Mid Water Trawl
(After Hameed and Boopendranath, 2000)

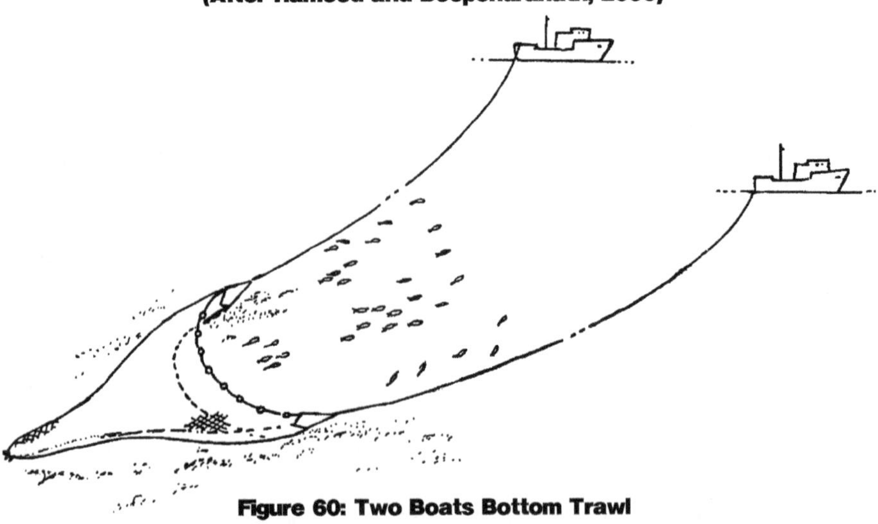

Figure 60: Two Boats Bottom Trawl
(After Hameed and Boopendranath, 2000)

Based on the method of keeping the mouth open, they are grouped into beam trawl (where mouth is kept open by means of a rigid wooden or steel beam fixed to the upper lip and remains open even if the boat stops (*Figure 62*) and otter trawl (where otter boards are used for horizontal spread of trawl mouth and the mouth

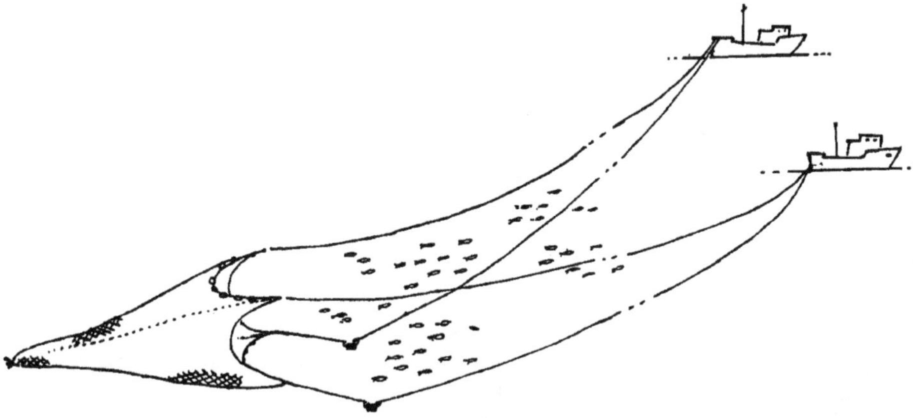

Figure 61: Two Boats Mid Water Trawl (After Hameed and Boopendranath, 2000)

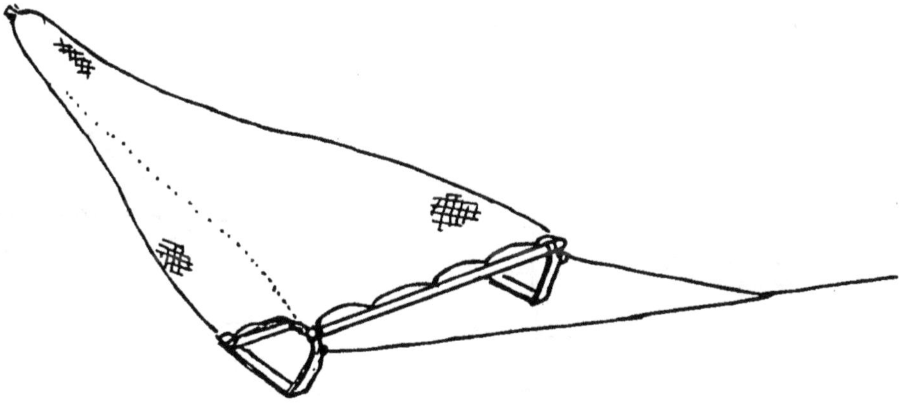

Figure 62: Beam Trawl (After Hameed and Boopendranath, 2000)

of net remains open only when boat is in motion (*Figure 60*). Based on number of trawls operated from single vessel, there are double rig trawl (where two nets are operated from outrigger booms (*Figure 63*), triple rig trawl (where three nets are operated at same time) (*Figure 64*) and quad rig system (where two nets are operated from two outrigger booms) (*Figure 65*).

Dredges

Dredges are dragged gear with oblong iron frame with an attached bag net (*Figure 66*). They are operated from the boat or in shallow waters by hand.

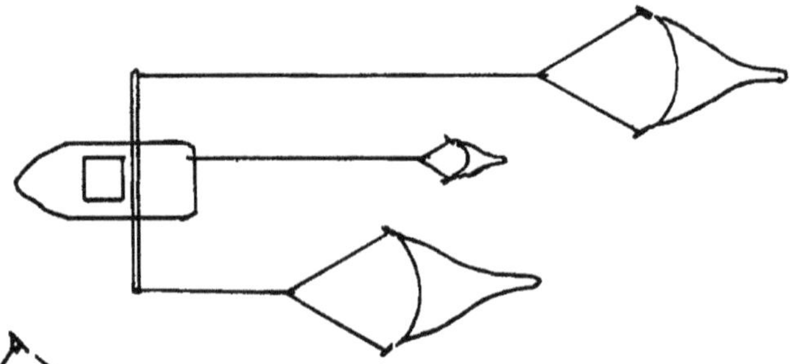

Figure 63: Double Rig Trawl

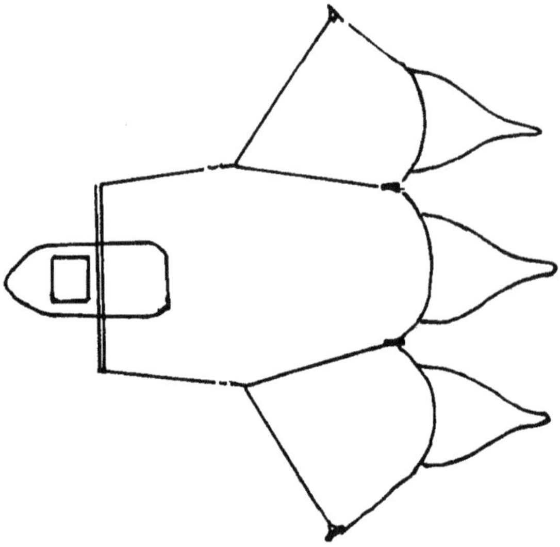

Figure 64: Triple Rig Trawl

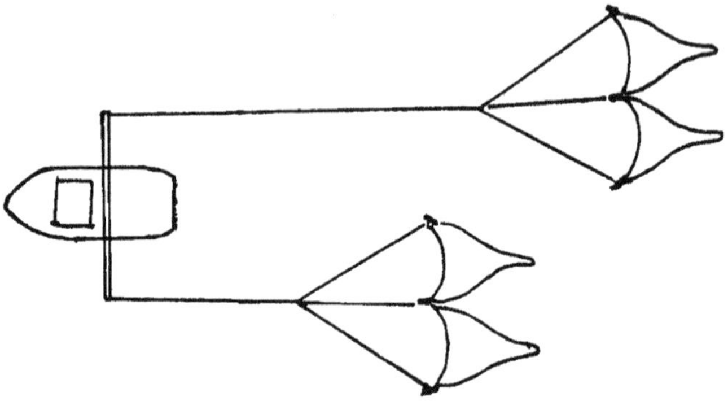

Figure 65: Quad Rig Trawl

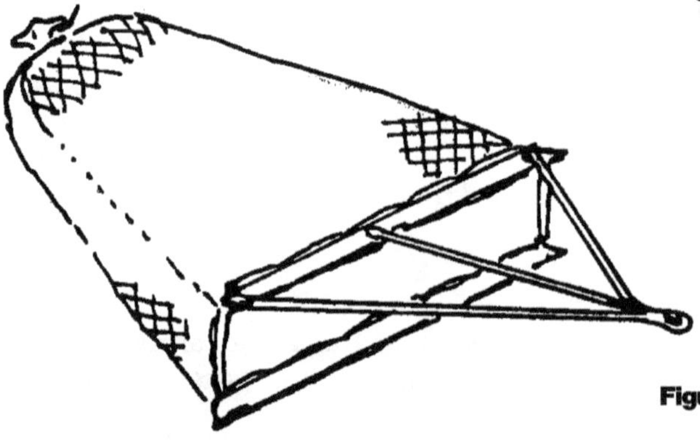

Figure 66: Dredge

Hooks and Lines

This method has been used since long. The principle of line fishing is to offer a real or artificial bait to entice the fish, which is then unable to release the bait and is lifted from the water. This gear is used to catch game fish as a part of sporting or to catch commercial fish (long lines). Previously thorn was used as hook, but now metallic hooks are used. The simplest form of this gear is "hand line" which consists of a cotton line (*dori*) with one or more hooks attached at one end and other hand is held in hand. This gear is very selective in action both in catching fish by species and by size. Nature of bait, size of bait and size of hook determines the selectivity of catch.

A fish hook is composed of a head or eye, shank, bend and point (*Figure 67*). Head serves for fastening the line and is shaped as 'eye' or 'loop'. The bend of work varies in shape. Long shank is specially designed to prevent the escape of fish after biting the line. The point of the hook may be provided with two or three barbs. A hook may be straight, bent or reversed, rounded, flattened, angular, double, treble and may or may not carry barbs.

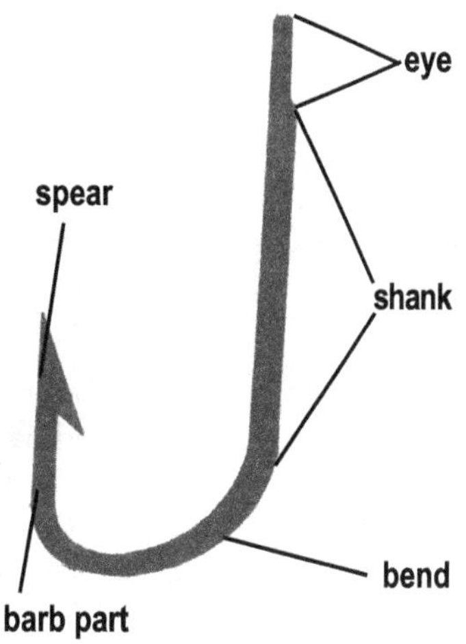

Figure 67: A Simple Hook

Bait is essential part and is selected carefully to attract fish. Live bait is most attractive but due to non-availability of it, dead bait is used. In hand lines, there is only one line with a sinker and snoods. Long lines now carry several hooks and are used to catch a large number of fish. The long line consists of a main line with several pendant vertical lines (*Figure 68*). Sometimes long lines are of great length and are provided with short branch lines or snoods bearing hooks, at equal distance. These lines can be operated very easily at places with high fish concentration. In most of the cases, long lines are shot from a slow moving boat. In muddy or rocky bottom, the long lines are held off the ground by floats. Long lines may be bottom set long lines (which set close to bottom), drifting long lines (when set in surface and mid water), vertical long lines (when set vertically) or bottom vertical long lines (with combining properties of bottom and vertical long lines).

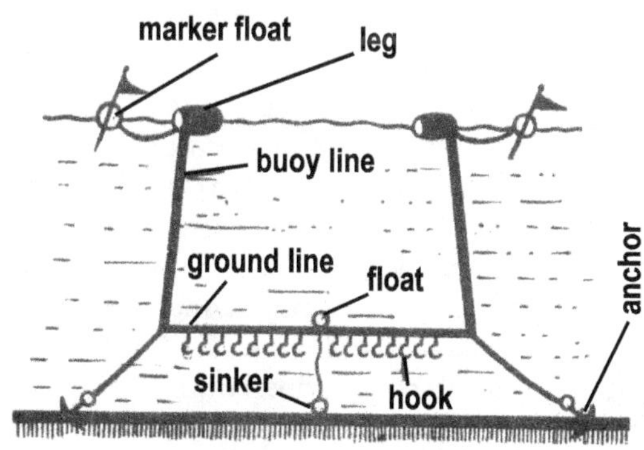

Figure 68: A Long Line

Important type of hooks and lines which are operated either manually or mechanically are pole and line (*Figure 69*); jig lines (which are operated either manually or by powered jigging machines for squids attracted by light) (*Figure 70*) and troll lines (for predatory fishes with hooks having baits, trailing behind the running vessel usually in surface layers (*Figure 71*).

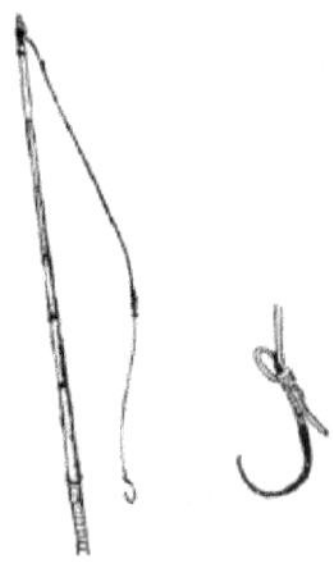

Figure 69: Pole and Line

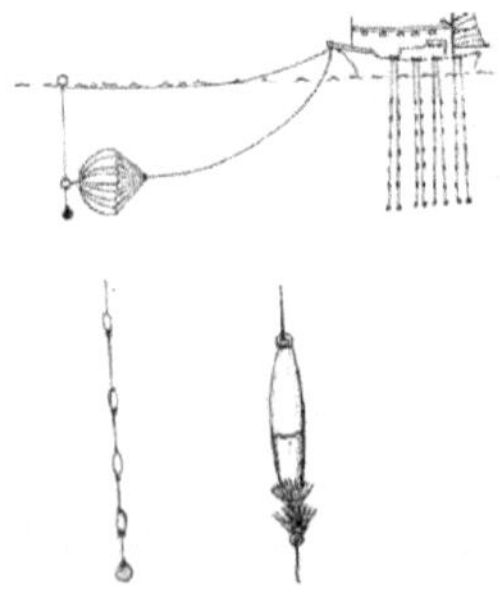

Figure 70: Squid Jigging

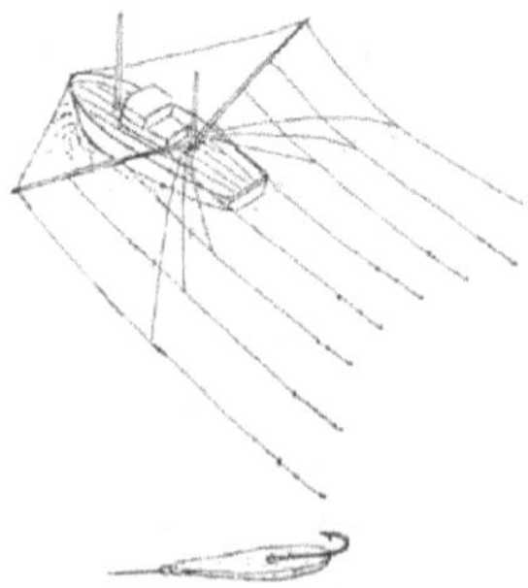

Figure 71: Troll Line

Lift Nets

Lift nets consist of horizontal netting panel or a cone-shaped bag with the mouth facing upwards, which are submerged and lifted either manually or mechanically to filter the fish in the overlying water column. Lift nets may be shore operated (*Figure 72*) and boat operated (when operated from one or several boats) (*Figure 73*).

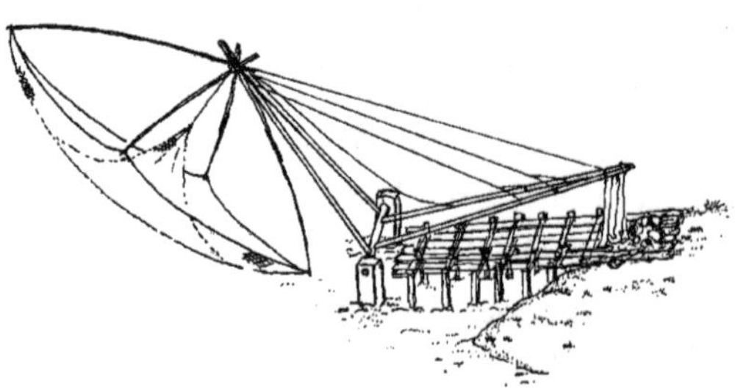

Figure 72: Shore Operated Lift Net

These nets are of varied shapes (triangular, rectangular or square) and are fitted round a frame. Smaller nets are operated by hand while the larger ones are dipped and lifted by means of long pole which is operated like a liver *(Figures 74 & 75)*. Lift nets can be described under four names:

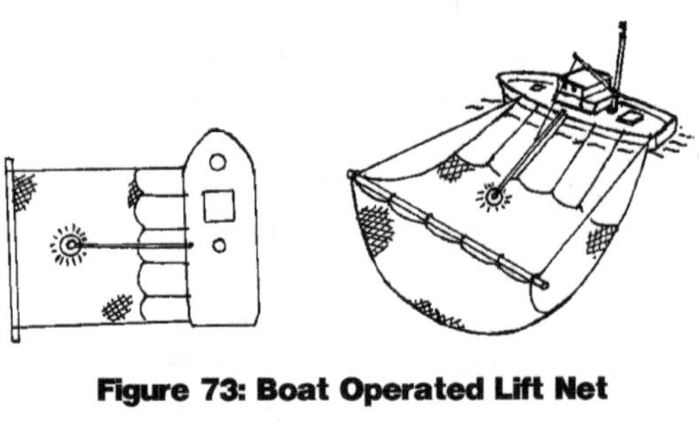

Figure 73: Boat Operated Lift Net

Hela Jal

This net is used to catch prawns catfishes and fry. It is a triangular net made of cotton and consists of two bamboos 2-5 m long crossing each other at one end to form 'V'. A cross stick about 50 cm long is fixed near apex of triangle and a bamboo pole is fixed to it to work as a handle. A net of 1.0–1.5 cm mesh is fixed to 'V'. During operation, broad end is pushed along the bottom and then lifted with a jerk to throw fish towards the handle.

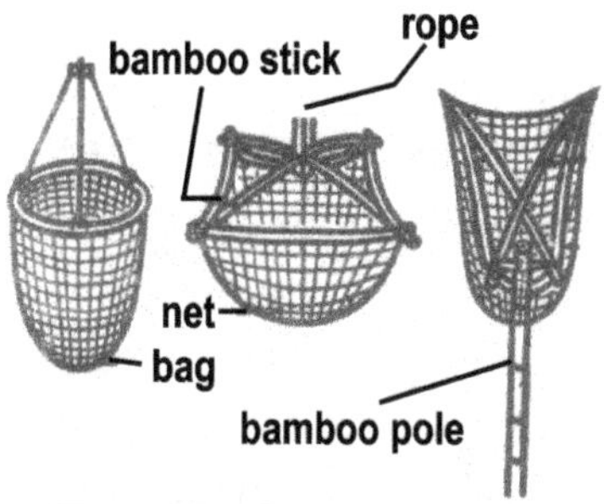

Figure 74: Smaller Nets

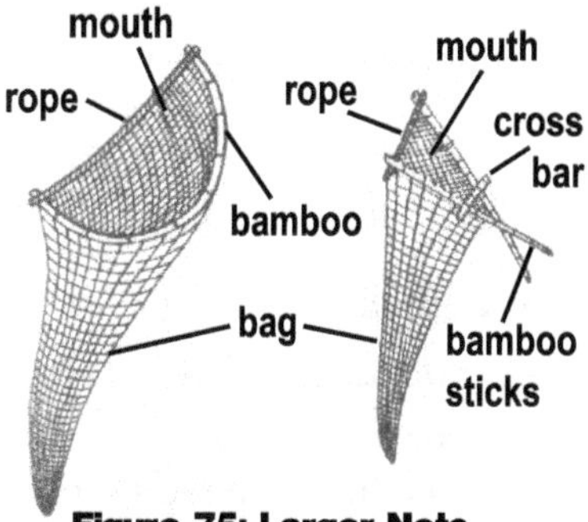

Figure 75: Larger Nets

Kharra Jal

It is a triangular large sized dip net constructed on the same pattern as Hela jal, but is operated from a platform made of bamboo pieces in river bed.

Bhesal Jal

It is similar to Kharra jal but is operated from a boat. The net is fixed to the V-shaped bamboo sticks and is allowed to hang like a bag.

Khoursla Jal

It is rectangular in shape with size 3 × 2 meter and is suspended from the ends of two curved bamboos crossing each other at right angles. The net is lowered in shallow water by means of a rope and is lifted suddenly when fish pass over it.

Falling Gear

Falling gear is cast over an area where fish is in abundance. It is then gathered and lifted to collect the fish. Cast net (*Figure 76*), cover pot (*Figure 77*), lantern net etc. are the examples of falling gear. The casting net is made up of a circular piece of netwebbing that is operated from boat. By holding the rope (hauling line) it is skillfully thrown over the water. The net spreads like umbrella over the shoal of fish. The perimeter starts sinking (being weighed with sinkers) and drags net towards

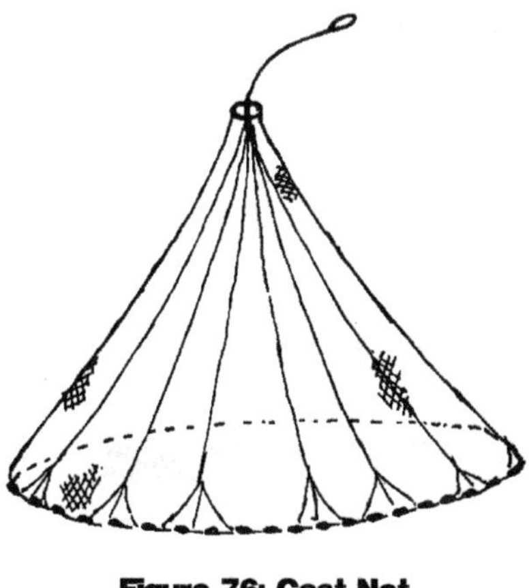

Figure 76: Cast Net

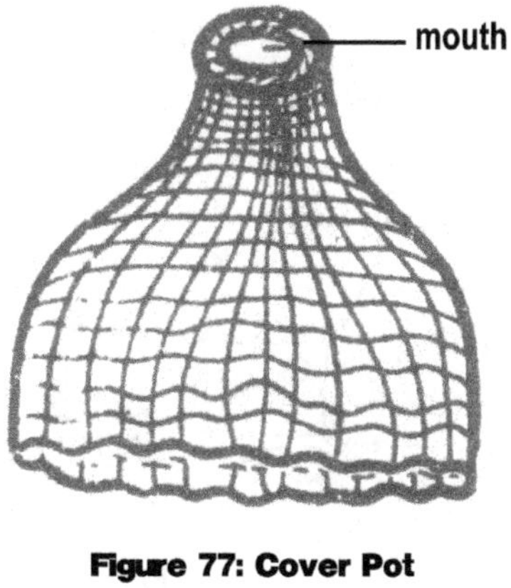

Figure 77: Cover Pot

bottom. The circumference is inwardly recurved so as to form an inner circular pocket around the perimeter and this prevents fish from escape during hauling when the perimeter touches the bottom, hauling line is pulled to raise the net on board and the trapped fish is collected. This net is used to catch small fish in shallow shore water or in river banks.

Passive Fishing Gears

Gill Nets and Entangling Nets

Gill nets are rectangular walls of netting kept erect by means of floats and sinkers and positioned in the swimming layer of the target fish, which catch the fish by holding them in the mesh by gilling. In entangled position fish is unable to disengage itself. The success in fishing is greatly enhanced by correct choice of the colour of the net. A colour that is invisible or less conspicuous to fish is likely to entangle more fish. Dark colour is good if water is limp, while dull or light colours if water is turbid. Nets even of nylon can also be dyed in any hue for camouflage. More fish are caught during night as fish is less likely to detect the net before it is caught so gill nets are generally left overnight and then hauled.

Depending upon the method of operation, gill nets may be drift gill nets (when operated in surface layers and drift with current) (*Figure 78*), set gill nets/anchored gill nets or fixed gill nets when fixed to the bottom or at a distance above bottom by means of anchors or ballast (*Figure 79*) and operated in shallow coastal water by means of stakes and catch is collected during low tide and encircling gill nets (operated in surface layer of coastal areas. After encircling fish, noise and other vibrations are used to drive the fish towards the net so that they are gilled or entangled).

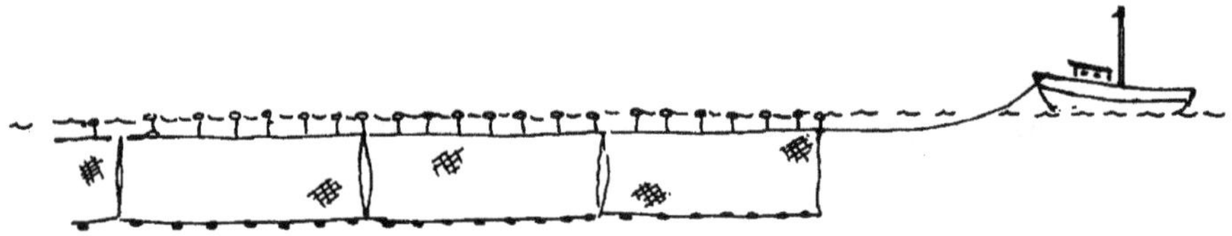

Figure 78: Drift Gill Net

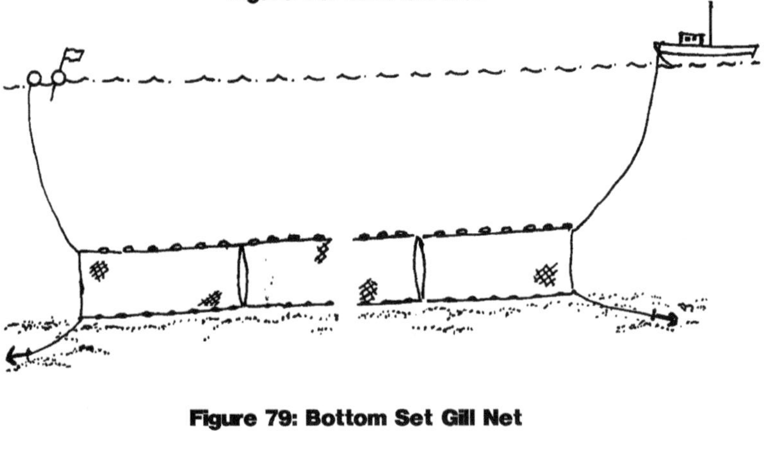

Figure 79: Bottom Set Gill Net

Gill nets on the basis of entangling of fish are of two types–simple gill nets and triple walled nets called trammel nets. Simple gill nets have a single wall of netting supported by floats and sinkers.

A loose net webbing is set in the path of fish movement. As a fish tries to swim through a mesh coming in its path, it is able to pass through the mesh by its head but not by its body which is too large for the mesh. In an attempt to withdraw its head, the twines slip over the gill cover and fish is entangled and cannot escape.

Such nets are highly sensitive in catching the fish, as a certain mesh size will gill fish only of a narrow size range. For capturing different fish sizes, different mesh sizes are required.

Gill nets may be set in three ways–floating, anchored and staked. Floating type of gill net is a wall of net with float line and lead line. The net is suspended and floats and weights are adjusted to maintain it either at surface or at a desired depth.

The anchored type is a modification of floating type. In this anchor used keeps the net stationary either at the surface or at any depth in mid water. In staked type the net is fixed at bottom with the help of anchors or stakes. Markers are held aloft on the surface to locate gill nets.

The triple walled nets or trammel nets have two outer walls of large mesh size and a loose inner wall of smaller mesh size (*Figure 80*). Trammel net catches wide range of fishes. In a combined gill net-trammel net, lower part is fabricated as trammel and upper part as simple gill net.

Figure 80: Trammel Net

Entangling nets are loosely hung single or multiple-walled netting held vertically in water by floats and sinkers. These nets catch fish by entangling rather than enmeshing. Nets are also attached end to end to form large fleets.

Fish Traps

A wide range of traditional fishing gears are grouped as set nets, pots or cages or baskets, fyke nets (used in shallow waters). Barriers, fences, weirs and corrals are trapping enclosures operated in tidal water.

Traps are used to lure fish which are allowed to get in but are prevented from escaping. The escape is made difficult by means of labyrinths or retarding devices like funnels or constrictions. These traps are grouped as:

A Basket Trap Net

The basket trap net (*Figure 81*) consists of two dome-shaped hemispherical baskets each with an opening at narrow end. This opening is guarded by flexible recurved bamboo sticks with their free ends facing towards the inner side. Bait in the form of balls is kept in trap and then lowered in water. Fish entering the basket are trapped and are unable to get out due to recurved nature of sticks guarding the opening.

Figure 81: Basket Trap

A Wide Mouthed Earthen Pot

Earthen pot or vessel is also used sometimes to trap fish. The mouth is closed with a thick cloth having a few holes to provide entrance. Bait placed inside the pot induces the fish to enter the pot.

Set Nets

These are stationary uncovered nets called set nets in Japan. A leader net is kept at an appropriate angle to the swimming direction of migrating fish schools so as to guide them to the enclosures with retarding devices and closed at bottom by netting (*Figure 82*).

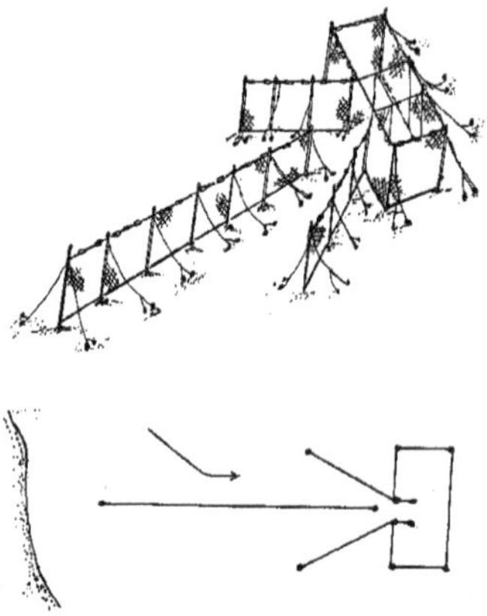

Figure 82: Set Net

Pots

These are cages or baskets made of wood, wicker, metal rods, wire netting etc. and designed to catch fish by enticing them with bait. The pots are provided with one or more entrances and usually set on the bottom singly or in series connected by ropes and position is marked by buoys (*Figure 83*).

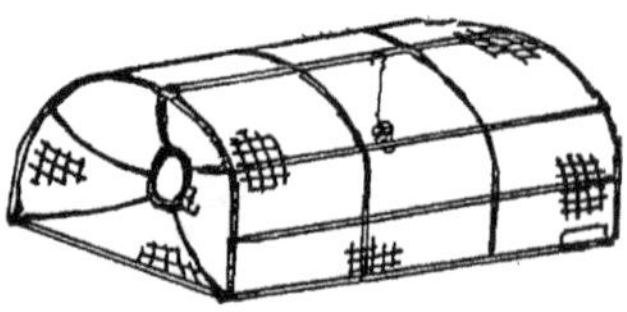

Figure 83: Lobster Pot

Fyke Nets

These are cone-shaped bags with ring-shaped rigid structure to maintain cylinderical shape of net body and is provided with wings to lead the fishes into the bag. The fyke nets are fixed to the bottom by stakes and are used in shallow water.

Stow Nets

These are conical bag nets operated in estuaries where tidal currents are strong (*Figure 84*).

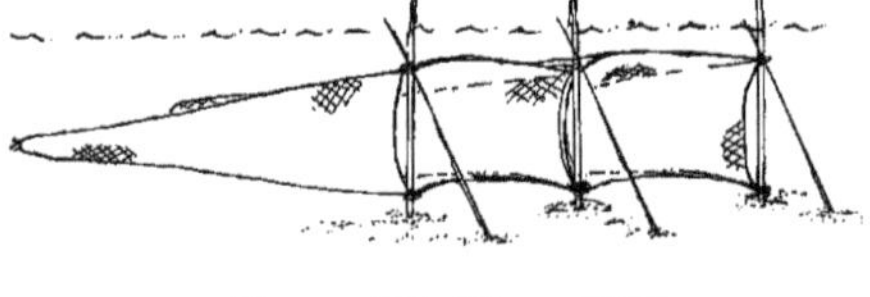

Figure 84: Stow Net

Barriers, Fences, Weirs and Corrals

These are trapping enclosures made of indigenous material and operated in tidal waters (*Figure 85*).

Aerial Traps

Fish that jumps out of water on disturbance or flying fishes attracted by light are caught in floating enclosures or rafts when they fall back. Examples are verandah net and boat operated aerial traps (*Figure 86*).

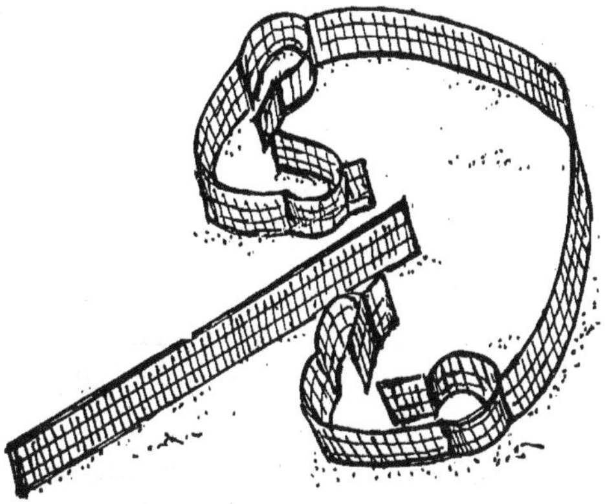

Figure 85: Corrals

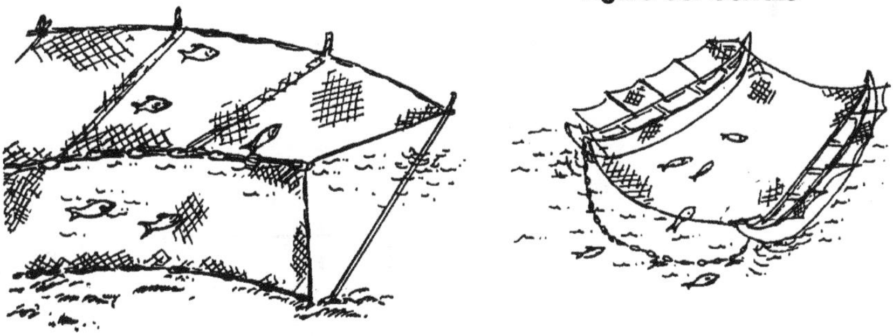

Figure 86: Aerial Traps

Hooks and Lines (Passively Operated)

Important hooks and lines which are passively operated are hand lines and long lines. The hand lines are operated in small scale sector while long lines have large number of hooks attached to it. Long lines when set close to bottom are called bottom-set long lines (*Figure 87*), when set in surface and mid water with freedom

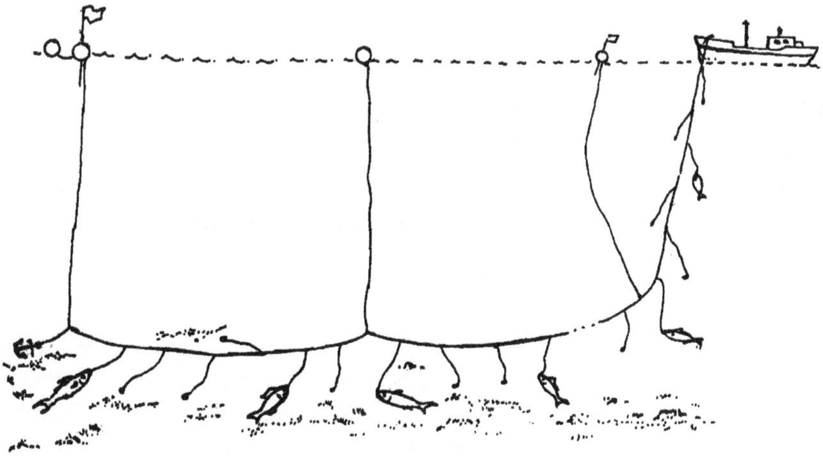

Figure 87: Set Long Line

to drift with current are called drift long lines (*Figure 88*), when set vertically are called as vertical long lines (*Figure 89*) and when possess the properties of bottom and vertical long lines they are called as bottom vertical long lines (*Figure 90*).

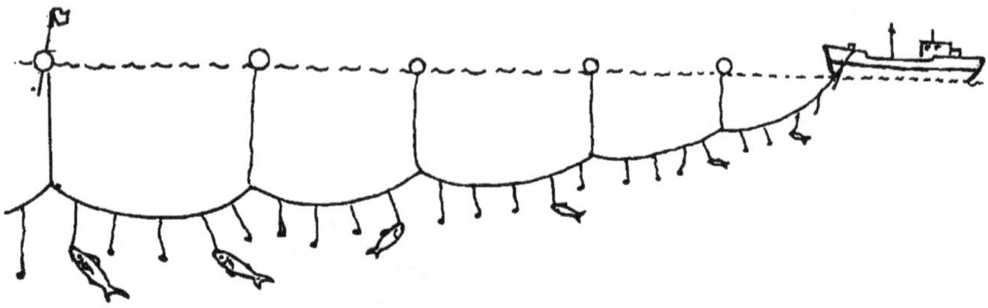

Figure 88: Drift Long Line

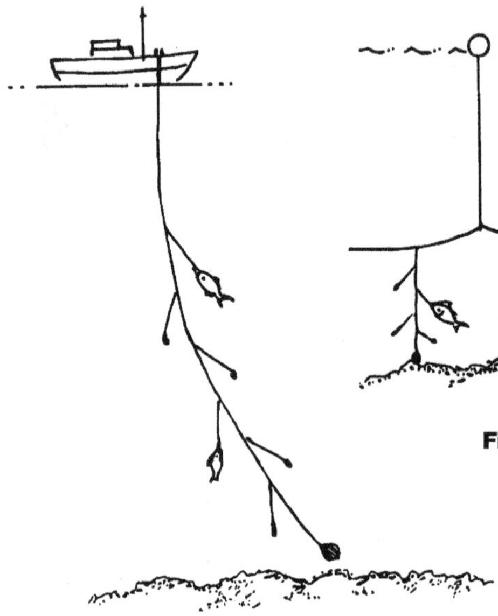

Figure 89: Vertical Long Line

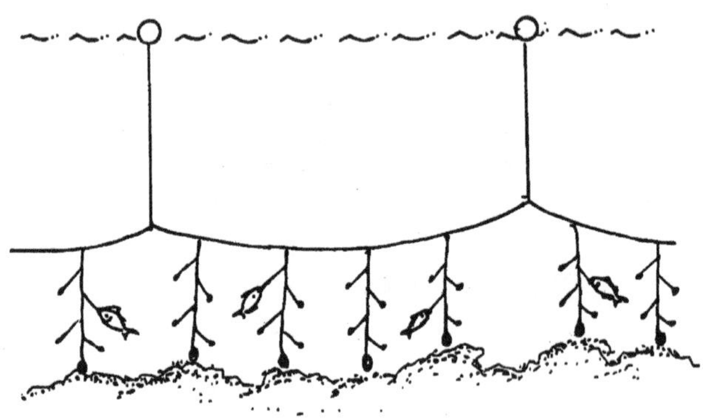

Figure 90: Bottom Vertical Long Line

Miscellaneous Fishing Gears

Fishing Without Gear

In this method, fish are hand-picked or caught by single implements as shovels, picks or knives, with or without support of diving equipment. Some trained pets are also used for capturing fishes, *e.g.*, cormorants.

Wounding Gear

Weapons like spear, harpoon, bow and arrow, lance and rifles are used to hunt large sized fishes. Ancient gear spear or 'konch' is used to kill large sized fish concentrated in an area. A spear consists of a split bamboo shaft 1.50-2 meters long fitted with a conical iron point. It is thrown by a person standing on the boat. Several modifications of the weapons are used under different names. "Harpoon" or "Ek-Katya" has a shaft of 3 meter long with one end having barbed iron point. "Rake" is also a special gear having long handle with long very thin blade at each end and looks like a double bladded oar.

Stupeflying Methods

It includes the use of poison or underwater explosives to paralyse the fish. This method indiscriminately kills the fish so scientifically is not very useful method. Certain chemicals affecting nervous system (central nervous system or peripheral nervous system) are also used. Rotenon obtained from *Derris eliptica, D. uligmosa* and *D. legensis* is a useful fish poison and is used in fish culture. The roots of this plant contains 10 per cent rotenon.

Electrical Fishing

It is a new metod of fishing by the use of electric current of low voltage. Since it has been observed that fish respond to polarity of electric field and swim towards the electrode with increasing potential. If two electrodes are put into water, fish starts moving towards positive pole. Thus a number of fish can be easily caught by placing anode in fishing net.

Harvesting Machines

Modern systems like fish pumps are used to mechanically transfer fish attracted and concentrated by light in the proximity of the vessel.

Various fishing gears used in rivers, hill streams, lagoons, backwaters and sea have been tabulated in the Tables 2 and 3.

Crafts

Fishing crafts in India are generally of primitive type. River crafts are light and flat as compared to larger ones used in sea. These crafts have been designed to deal with different kinds of fishing/fishing gears and are built according to local conditions.

Rafts and **dongas** are the primitive types of crafts. They are operated in calm waters. **Plank-built boats** are used in large rivers and estuaries.

Rafts made of different material like inflated buffalo skins tied together are generally found in upper reaches of river Ganges. Banana stems or shola bundles are tied together to form floating platform.

Several inverted earthernware pots are laced together in a bamboo frame work and are also used as a platform in some marshy areas, ponds and jheels.

Dug-out canoes called **Dongas** used in Bengal, **Ekhta** in Bihar, **Odam** in Malabar are commonly used crafts in shallow waters of tributaries. **Vallam** is used in backwaters and estuaries of Kerala. **Chatty rafts** (earthen pot rafts) are commonly used in Tanjore district (Chennai), in river Cauvery in Tiruchirapalli. The **Coracle** used in rivers Cauvery and Tungbhadra is also used in some reservoirs. **Tigarni** or **gamla** similar to Coracle is used as craft in West Bengal.

Table 2: Gears Used in Inland Freshwater, Lagoons and Backwaters

Gears Used in Rivers

Sl.No.	*Gear types*	*Characteristics*	*Area Operatable*	*Method of Operation*	*Remarks*
1.	**Seines**				
	a. Berjals	Different dimensions and different mesh size are used	Gangetic area	—	**These jals have a stout foot rope, absence of bag at mid length and no sinkers. Net may be loaded upon more than one boat and then hauled on shore by several fishermen**
	b. Kona jal or Bhasa gulli	Size 91 m × 9 m cotton net	Used in Hilsa fishing	Small meshed pockets, conical in shape are attached along net to catch fish. Due to valves in these pockets fish cannot escape and thus captured	—
	c. Jagat ber or Mahajal	—	Riverine area	—	**Captured throughout year except July to September**
2.	**Drag net**				
	a. Moi or Moiajal		Shallow water	—	**Can be operated by two fishermen**
	b. Chunti jal		Bihar		
	c. Mahajal (long drag net)	Size 4 × 2 m	Rivers with strong current	—	**Huge amount of fishes are captured**
3.	**Drift net**				
	a. Keral or Katla jal	Carp and Katla capture	Narmada river	—	**It consists of two bamboo poles crossing each other near two long sides**

Contd...

Table 2–Contd...

Sl.No.	Gear types	Characteristics	Area Operatable	Method of Operation	Remarks
4.	**Dip net**				
	a. Mela jal	Catfish capture	Narmada river	—	**It is like Hela jal but is operated from bamboo stage**
	b. Karra jal	Hilsa capture resembles open bag			**Bhil fishermen catch Hilsa by this net**
	c. Jamda	Stationary net with mouth kept open by vertical rod			
	d. Suti jal				
5.	**Purse nets**				
	a. Kharki jal	Hilsa capture	—	—	**In this jal a vertical bamboo rod is attached to lower margin of mouth to open and close it**
	b. Shanglo jal or Skarki		Upper reaches of estuaries		**In this purse is opened and closed with weighted cord**
6.	**Cast net**				
	a. Bechari or Otter jal	Catla, Rohu capture (area more than 55 m)	—	—	**Since peripheral area is large, so 3-4 fishermen operate it from boat.**
7.	**Kuriar net**	Catches big carps and Hilsa	Used in shallow waters	Net is kept inverted and dipped and then taken out of water	**This net can be operated single handed**
8.	**Bamboo Screens**	—	Shallow ponds and small rivers	Manipulated and handled like seine	**Bamboo screen and drip net are combined sometimes for operation**
9.	**Fixed nets**	—	Used in rivers and tributaries of Uttar Pradesh	Long rectangular piece of mesh stretched from one bank to another bank and held up by bamboo sticks in the middle.	—

Contd...

Table 2–Contd...

Sl.No.	Gear types	Characteristics	Area Operatable	Method of Operation	Remarks
10.	**Triangular net commonly known as 'Bhog'**	Conical net shaped like butterfly and made up of strings	Eastern districts of Uttar Pradesh	Mouth of net is kept open by bamboo sticks. One of the sticks is longer and used as handle. The closed end of the net serves as reservoir for collection of fish.	**The net may be operated from boat also**
11.	**Trap nets**	Commonly used	Uttar Pradesh		
	a. Basket trap net	Consists of 2 dome shaped baskets with suitable bait in the form of ball is placed in it	–do–	The opening is guarded by bamboo sticks and fishes entering bakset are trapable to come out due to recurved natue of sticks guarding the opening	—
	b. Pot trap net	Commonly used by poor people.	Eastern Uttar Pradesh.	Wide mouthed earthen pot is used as trap with a thick cloth having few holes and suitable bait put inside the pot tempts fish to enter	**Live fishes like Channa, Clarias, Heteropneustes inside pot are captured by hand**
	c. Konch trap net	Used in shallow muddy waters of summer	—	Made of split bamboo pieces in form of conical basket with small circular opening at the top to allow the hand to enter. The trap is dropped in water and the wide mouth is pressed in soft mud. The fishermen then brings his hand through the top opening and catches the fishes which wriggle in muddy shallow water	**Channa, Heteropneustes, Clarias, Cuchia, eels are collected by this method**
12	**Electric equipment**	Current is provided by generator or battery	Ponds	Fish with electric field receive shock and they are made numb for a few moments and during this period they are seized	—

Table 2–Contd...

Gears Used in Hill Streams

Sl.No.	Gear types	Characteristics	Area Operatable	Method of Operation	Remarks
1.	**Seines**	Commonly used	—	—	—
2.	**Dragnets**	Commonly used	—	—	—
3.	**Basket trap**	Some part of stream is bunded off	Stream parts	—	—
4.	**Explosives & Poisons**	To capture on large scale	—	—	—

Gears Used in Lagoons and Backwaters

Sl.No.	Gear types	Characteristics	Area Operatable	Method of Operation	Remarks
1.	**Bag nets**	Used against tidal currents		Some area is enclosed by bamboo screen and then by draining it fish are caught	—
	a. Cast net or Dip net		Sunderbans		
	b. Khalpaltta jal		Sunderbans		
	c. Charpatta jal		Sunderbans		
	d. Thatturala		Thatturala (Kerala)		
	e. Janos		Chilka		
2.	Bush netting	—	Sunderbans	Bamboo screens reused to enclose an area and that area is piled with bush to attract fish	—
3	**Seine net**	Having 4-5 boats and 24 fishermen		—	—
	a. Kochal jal		Sunderbans		
	b. Jangle jal		Sunderbans		

Contd...

Table 2–Contd...

Sl.No.	Gear types	Characteristics	Area Operatable	Method of Operation	Remarks
4	**Drag net**				
	a. Mori jal	—	Chilka Lake	—	—
	b. Khadi jal		Chilka Lake		
	c. Sahalo jal		Chilka Lake		
	d. Patua jal		Chilka Lake		
	e. Mani jal		Chilka Lake		
	f. Ber jal		Chilka Lake		
	g. Jagal jal		Chilka Lake		
5	**Purse net**				
	a. Shanglo jal	To capture large fishes	Estuarine area of Ganges	—	—
	b. Kharki jal		Upper areas of estuaries		
6	**Drift gill net**				
	a. Chandni jal	More than 300 m long and 3 m wide	Estuarine area of W. Bengal	—	—
7	**Gill nets**				
	a. Menjia jal	—	Chilka Lake	—	—
	b. Noli jal		Chilka Lake		
8	**Rangoon nets**	Hilsa capture	Estuaries of Krishna and Godavari	—	—
9	**Chinese dip net**				
	a. Cheenavala or Kambuvale	9 – 11 m size.	Backwaters of Travancore and Cochin	—	—
	b. Changatopayilal or Changadem	Mullet fishing	– do –		
	c. Rod and Line Fishing and Handling		Backwaters of Kerala		

Table 3: Gears Used in Sea

Sl.No.	Gear types	Characteristics	Area Operatable	Method of Operation	Remarks
1.	**Fixed nets**	Conical or rectangle shaped with various size and meshes	Sunderbans Sunderbans Sunderbans Thatturala (Kerala) Chilka	Some area is enclosed by bamboo screen and then by draining it fish are caught	—
	Panch Kathia-Kool jal Behudi or Ghurni jal & Panch Kathrabai jal	Conical nets	W. Bengal & Orissa		
	Jadi or Mtagh jal		Gujarat & Kathiawar coast		
	Kalem-Kettavalai Bayd or Mal jal	Rectangular nets	Gulf of Munar, W. Bengal and Orissa		
	Barnade jal Kalavalai		N. Orissa Tanjore district Kanara coast		
	Kondavala, Wagholjal				
2	**Seines** (Large nets)	Active fishing	Seashore operated	—	—
	a. Shore seine				
	b. Boat seine				
	c. Beach seine				
	d. Danish seine		Used in deep water		
	e. Purse seine				
3	**Scoop nets** (Dip nets)	Round shaped net with mesh proportionate to size of fish	—	Net is moved swiftly in a scooping manner to lift fish out of water	—
4	**Trap nets and Nets**		Used in shallow water	—	

Contd...

Table 3–Contd...

Sl.No.	Gear types	Characteristics	Area Operatable	Method of Operation	Remarks
	Fyke nets or Loop nets	Used to catch catfish and pike spawners			**Large type of trap nets**
	Pound nets		–do–		
5	**Drop net**	—	—	Net is mounted squarely and suspended at its corners by supple hoops which are tied at top and attached to pole. It can be pulled up and down. When drop net is pulled up, net forms deep pocket which holds the fish	**Its dimension vary**
6	**Cast net**	Hand operated with string or without string	—	—	—
7	**Drift net & Gill nets**	Nets of entangling type	—	—	—
	a. Simple gill nets				
	b. Trammel net				
8	**Hooks & Lines**	Baited hooks are used	—	—	—
	a. Hand lines				
	b. Long lines				
9	**Trawl**	Large towed nets.	—	Catch fish by filtering water	—
	a. Otter trawls				
	b. Beam trawls				

Plank-built boats are used in rivers with strong currents and tides. Small riverine and estuarine crafts are called **dinghis**. They are used in West Bengal for operation of purse seine and dip nets. Dinghis have narrow tapering bows and sterns but they have no keels. The **Chhandi nauka** is used for operation of drift nets. **Bachari** is another type of large boat like Chhandi nauka. It is slightly longer but narrower than Chhandi boat.

Mechho bachari is a medium sized boat used for transporting live fish in Kolkata (erstwhile Calcutta in West Bengal).

In Chilka lake and Mahanadi, **nava**, a plank-built boat is used. **Machwa** type boat is used for operation of large nets in estuaries of Gujarat.

Khulnawa–It is a special type of craft used in Ganges. Khulnawa is a local name which means an open wale boat. It is said to have been introduced from Tarai area of Nepal. This is operated at night in calm weather from February to April. It has three components in its structures:

1. Boat with one of the wales open;
2. Screen platform; and
3. A frilled pole.

In its construction, Kathal or Sal wood is used. It is 7.5 m in length. The wall on one side has a cut opening which is 5 m long and 12 cm deep. The boat is darkened by coaltar coating.

The screen platform is closely knit screen of bamboo splinters measuring 5 m × 0.5 m. The screen is supported and reinforced by broader splinters of bamboo 70 cm long at an interval of 50 cm. Over the width of screen, extra length of splinters project inwards. Here they are attached to the boat providing room for fastening, with strings or pegs which are fixed on the floor of the boat. The free margins of screen are constantly immersed in water and always kept sparkling white by coating of enamel paint.

The frilled pole has a piece of 4-6 m long bamboo pole with tufts of dry grass (*Pollinidium anguistifolium*) dropping as frill along 3/4th of its length. Series of tufts of grass are wrought to make a dropping frill along the length of the pole. The whole frill is tied at intervals (50-60 cm) and it hangs from the pole. Khulnawa is operated by two fishermen. Side of boat with open wale and screen faces the bank of river. The free edge of dropping grass is kept overhanging the surface of water. In the dark the sparkling light of screen platform gives the fishes an illusion of flowing stream. Minnows happen to be schooling in between the boat and the bank.

Sea Fishing Crafts

Following are important sea fishing crafts:

Catamaran

It is the most rudimentary craft used by any group of traditional fishermen all over the world. The word "catamaran" is derived from Tamil "Kattumaran" which describes the nature of craft. It is a keel-less craft formed by lashing together 3-7 pieces of light rough-hewn wood. A loose rubber is hung between the logs. It is known as "teppa". Catamarans have a length of 5-12 m, width 0.7-1.4 m and depth 0.3-0.7 m. It is constructed with logs of wood.

Generally, catamaran consists of two main logs and two side logs. They are cut into boat-shape and tied together with rope. There are four types of catamarans:

Orissa and Ganjam Type

These are boat-shaped and made up of five logs pegged with wood instead of being tied together.

Andhra Type

It is slightly larger than Orissa and Ganjam type. It has heavy wood. These are strong median logs used in fitting the sides.

Coromandel Type

It is made up of 3-5 logs with variations in pattern. It is commonly used in Madras and in flying fishery of Nagapanam Kolamaram 7-logged catamaran is used.

Boat Catamaran

It is used on coast around Mandapam. It has three logs fitted into a regular boat shape. Variations in this type occur in Cape Comorin, Tuticorin and other areas.

Masula Boat

It is keel-less boat constructed from mango planks stitched together with palm leaf fibres. It is operated near shore. These boats are 8-12 m in length and are of weak construction. It can be used in fair weather only. The small masula boats are extensively used in gill fish netting. Examples are Barboats of Orissa and Padava or Padagu in Andhra coast.

Dinghi and Nauka

These are commonly operated in West Bengal and Orissa. Naukas (12–13 m × 3 m × 2m) are also used for fishing and other purposes (*Figure 91*).

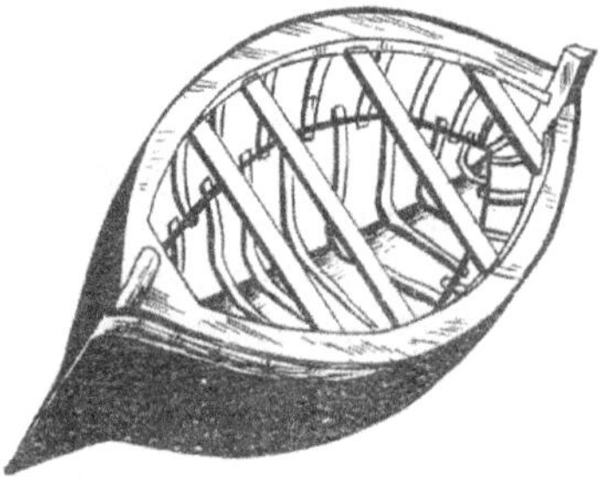

Figure 91: Dinghi

Tuticorin Boats

These are called fishing luggers. These carvel boats upto a size of 10–11 m × 2 m × 1 m and are operated in inshore water. They are also used as motherships and cargo boats (*Figure 92*).

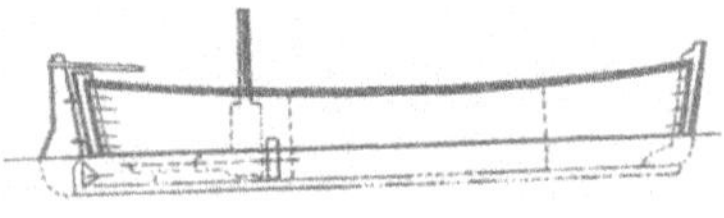

Figure 92: Tuticorin Boat

Dugout Canoes

These are used on Kerala and Canara coasts. These are made from hollow trunk of Palmyran tree. The large dug out canoes comprises of main fishing craft of Malabar coast. These are 10-20 m long boats which can be operated by 1 or 2 fishermen and can be used for fishing in rivers and canals.

Plant Built Canoes

Here dugouts are further enlarged with planks on the sides. These are used in Kerala, north Mumbai and Kathiawar.

Outrigger Canoes

These are large (upto 17 m long) in size with narrow keel and more spread out planks. Canoes with single outrigger are used on Canara and Konkan coast, also called Rampani boats because they are used for mackerel operations with rampani nets.

Built-up Boats

Can be seen on west coasts along Mumbai-Cambay coast and north of Ratnagiri. It has a pointed bow, straight and narrow keel and low gunwale. This is particularly known as Ratnagiri type of boat. The Bassein type having broad hull, pointed bow and straight keel is locally known as Machwa. The Satpati type known as Galbat, has a broad beam, median pointed bow and straight keel and high gunwale (*Figure 93*). The Broach type is flat bottomed and is of great use in inshore and estuarine water. The Satpati type of built-up boat can be easily mechanised.

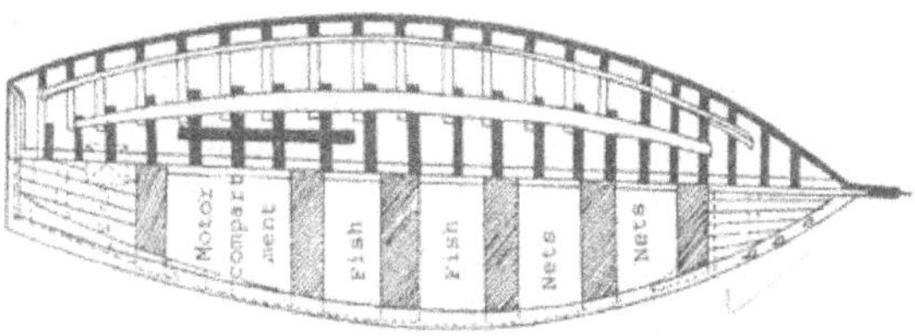

Figure 93: Satpati Type Fishing Boat

Shoe Dhonie

Craft resembles shoe and can be used in sea and rivers. It is common in Godavari. It is made of teak wood and is constructed with planks grooved with ribs and frames. Nails are also used. Three to four persons can operate it for fishing with gill nets.

Coracle

It can be operated in rivers, canals and reservoirs with two to three fishermen. Its frame is constructed with split bamboos, the outer side being covered with leather.

Kakinada Nava

The craft is very common for inshore fishing. It is keel-less boat of 7.5 m length, made of teak wood and carvel built (*Figure 94*). They are constructed with frames and ribs. Nails are also used in their construction. It is of open type and can be used in lakes, rivers and reservoirs.

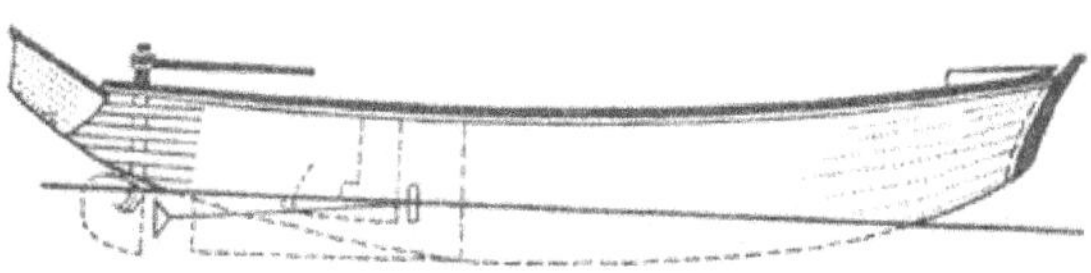

Figure 94: Kakinada Nava (Andhra Pradesh)

Unconventional Fishing Methods

These are of recent origin but have high prospects. They include:

1. Electrofishing (using electric current).
2. Light fishing (using artificial light).
3. Fish finder (echo sounder, sonar, fish magnifier using sound waves).

Generally, cast nets are used along river margins and not in deep waters. Drag nets, gill nets or scoop nets are used in tanks, pools, reservoirs, where water is fairly deep. Some game fishes like mahaseer and some catfishes are caught by hook and line. Hill stream fishes are often collected by damming a stretch of river and/or diverting the water stream to dry up. Fishes are then picked up by hand from below the stones, rock and boulders. Some roots of plants like *Derris* sp. nacrotises such hill stream fishes and they are easily picked up. Traps are also used to collect some eel-like fishes.

Mechanization of Indian Crafts

Fisheries sector is predominated by traditional crafts and gears. Nets are made of natural fibres mainly cotton and hemp. But now, mechanised boats have been introduced in large number. Durable nylon yarn is used for this purpose. For increased production, craft motorization and production of specialised boats seemed to be essential. Diesel engines are now obtained and are utilised for motorization of *navas* and even catamarans. Various states are doing their best for improvement of fishing crafts and gears.

Modernization of Fishing Methods

In India, fishing gears and crafts used to catch fish are primitive due to which considerable time is wasted in reaching fishing grounds and fishermen return with only a small catch. Now attention is being paid towards modernization.

//••••••••••//

Chapter 4

Sorting, Preservation and Measurement

Sorting

Once the fish are collected and brought to the laboratory, these need to be sorted out on the basis of their shapes and close similarities into lots. Each lot should be labelled indicating the name of the locality *viz.*, river/pond/tank etc., distance from the town/village, date, time, method of collection (mention net, trap or poison etc.) and name of the collector.

During collection all possible geographical information must be gathered. An analysis of fish habitat and its environment should be made and recorded. At the collection site, the physico-chemical qualities of water (colour, water depth, temperature, shore line, type of bottom etc.) and the type of methods for fish collection should also be recorded.

There is always a need to record superficial characters of the fish, as the same may get distorted when preserved in formalin. The features of importance for visual recognition are colour of the fish (of special interest are spots, blotches, numbers and designs of bands or stripes on live fish) and position of lateral line. There is always a need to define and reveal the local name of the fish. This is possible through fishermen or people living in the vicinity.

Preservation

Fishes caught alive or in narcotised state are put in a dilute formalin 10 per cent solution. The container should be covered with lid or cloth so that fish may not jump out. The fish should remain immersed in formalin solution for about 4-5 hours so that these may become hard and well-preserved. Formalin being acidic softens the bony structures. The addition of household borax to formalin solution retards the shrinkage of specimens and softening of bony tissues. Fish preserved in formalin die slowly by expanding its fins and rays which will help in identification. Marine fishes can be preserved in 2-4 per cent formalin solution.

Fish less than 10 cm long could be preserved by immersing in formalin completely and fish of length more than 10 cms (upto one foot) need to be incised in the abdominal wall, a little to one side of mid-ventral line with a scalpel, knife or scissors or slit between the pectoral and pelvic fin bases on the right side to enable the formalin to preserve internal organs. Deep cuts, may damage the alimentary canal, therefore should be avoided. Fish longer than 30 cm should be injected undiluted formalin in several places along the abdomen with the help of hypodermic syringe and needle. Belly should be slit at several places with 2-5 cm long incisions. Where abdomen is not round but sharp and keeled, make incisions on left side of the fish. Fishes can also be preserved in alcohol. They should be kept in formalin for 4 days after which they are washed in tap water for 3-4 hours and then stored in 75 per cent ethyl alcohol. Fishes that are too long to be stored in a straight position are not bent or coiled but are cut either in half or through 9/10th of the body and then placed vertically in a storage containers. When the fishes are slit open for preservation, intestinal bulb comes out. This bulb could also be preserved along with fish for food studies and subsequent analysis regarding their habitat. Many workers consider it beneficial to remove the slime adhering to the scales before preservation, it can be rubbed off using a dry cloth or a cloth soaked in a dilute solution of borax or alum in tap water. Ethanol and isopropanol are used to preserve skeletal structures like otoliths etc.

For transport of fish to short distances, collection can be kept in plastic bags covered with formalin soaked cloth or cotton. For transport to distant places they must be packed in the mulmul cloth and sent by air, rail or road. The container should be clean and leak proof.

Colour Preservation

Waller and Eschmeyer (1965) suggested a chemical antioxidant, butylated hydroxytoulene (BHT), in a liquid emulsifiable concentrate to preserve colour in fishes. Orange and red coloured fishes when treated with BHT with a concentration of 20 cm^3/4500 cm^3 of 10 per cent formalin retain their colour after 18 months. Red, orange, yellow, brown and black coloured fishes when tested, did not lose their colour.

The colour preservation only works when the storage containers are filled completely with the solutions to exclude air. Oxidation will occur if air is allowed to contact the specimen, and this will cause the colour to fade.

Measurements

Sharp pointed needle-like dividers or a diel reading calipers are used for taking body measurements. Dividers should have one point flat at right angle to the place of operation and other kept at a needle point. For accurate readings, a stainless steel ruler with measurements to millimeter is recommended.

Unit of Measurement

It is recommended that the metric system be adopted universally for all scientific measurements relating to the assessment of fish stock.

Measuring linear dimensions of whole or part of fish is probably the most widely used technique in fishery biology studies. Such observations are made with special measuring boards, tapes and calipers. Measurements are usually but not always taken along straight lines.

The measurement of fish over 10 cm in length should not normally be quoted to millimeter precision. For fish ranging from 10-50 cm in length, quarter or half centimeter intervals may be satisfactory.

The offset board has some disadvantages. It cannot be used easily as a scale for other purposes unless it has a removable head piece block one half unit thick.

Design of Measuring Boards

Here are given some general principles of design, illustrated with reference to the requirement for field measurements of total length of clupeoids, carangids and scombroids not exceeding 40 cm. For this purpose a half-meter scale would usually be adequate. The board would, however, be used to measure other kinds of fish. An extension piece, or hinged section would permit measurements upto one meter.

If measurement of standard length and other dimensions or overall length are to be measured, it may be necessary to fit a sliding cursor carrying a wire which passes over the fish when it is lying on the board. Such boards sometimes also have a mirror at the base with which parallax error is avoided. Devices of this kind are more often used in the laboratory for morphometric studies than for routine observations in the field.

Length Measurement of Fish

Length measurements are usually made with the fish lying on its right side, snout to the left, on the measuring board consisting essentially of a wooden or metal base carrying a central scale and having a head piece (nose block) against which the snout is generally pressed (*Figure 95*). The mouth is closed, the fish body and tail straightened along the mid-line and the readings are taken from scale. An attempt is usually made to measure the whole fish in the fresh wet condition; if *rigor mortis* has set in, fish should be gently fixed just before measuring. In certain circumstances (during tagging experiments), when fish need to be measured alive, special steps like injection of narcotics or relaxant should be done before handling of fishes.

Large flat fishes present difficulties. Their length may be measured with calipers or from point-to-point along the body surface with tape.

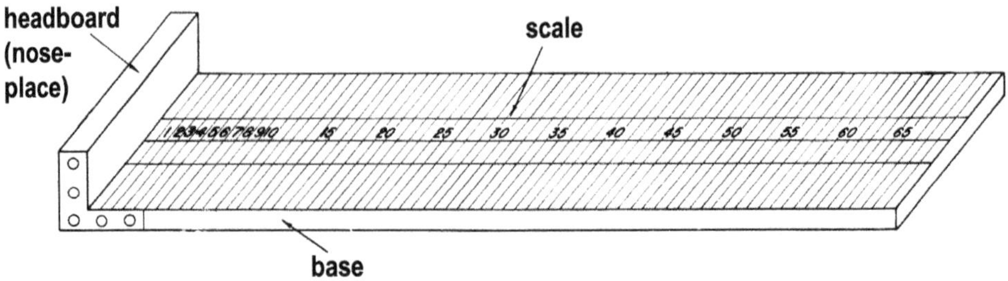

Figure 95: Measuring Board

Rays and other dorsoventrally flattened fish may be measured while laying straight on ventral surface; the standardised linear dimension of rays is often the disk width rather than overall length.

There are many more measurements which can be taken from a fish. However, recording of the data is very essential and it needs to be understood that without an appropriate collection data, a fish like other scientific specimen is completely useless for research. A typical data sheet recommended by various authors need to be filled-up for each specimen or for each batch of specimens. Individual specimens collected at different localities receive individual number and a data sheet. This data sheet can be maintained as raised by Australian Museum (Hangay and Dingley, 1965) (*Figure 96*) or placed in Manuals of Methods of Fisheries Biology (T. Laevastu, 1965) (Table 4).

AUSTRALIAN MUSEUM - DEPARTMENT OF ICHTHYOLOGY - REGISTER CODING SHEET

COMPLETE _______ CHECKED _______

PAGE 1 OF _______ NO. _______

PUNCHING INSTRUCTIONS :

1. Punch only lines numbered in column 1.
2. Lines 5, 6 and 7 to be ignored if columns 2 - 80 are blank.
3. Lines with 6 & 7 in columns 1 to be ignored if columns 1 - 80 are blank.

ELD NUMBER _______________

ARDS TYPED _________ NOT TYPED _________

DATE REGISTERED _______________

REGISTERED BY _______________

Columns: 1 2 3 4 5 6 7 8 9 10 11 12 13 14 15 16 17 18 19 20 21 22 23 24 25 26 27 28 29 30 31 32 33 34 35 36 37 38 39 40 41 42 43 44 45 46 47 48 49 50 51 52 53 54 55 56 57 58 59 60 61 62 63 64 65 66 67 68 69 70 71 72 73 74 75 76 77 78 79 80

Line	Col. 1	Fields
1	1 1	REG. NO. / COUNTRY, STATE OR OCEAN / GENERAL LOCALITY OR DRAINAGE SYSTEM / SPECIFIC LOCALITY
2	2	COLL. DATE / CAPTURE - DEPTH / WATER - DEPTH / TIME - (RANGE) / COLLECTOR / FIELD NUMBER
3	3	BOTTOM
4	4	LOTS / LAT. / MIN. / LONG / MIN. / F / FORMER REGISTRATION NUMBER
5	5	
6	5	
7	5	
(6)		SP. NO. / FAM / GENUS / SPECIFIC NAME / NUMBER / SIZE / RANGE / T / A / S
8	6	
9	7	
10	8	
11	7	
12	6	
13	7	
14	6	
15	7	
16	6	
17	7	
18	6	
19	7	
(7)		IDENTIFIED BY / DATE / REMARKS : AUTHOR, DATE, SUBSPECIES ETC.

GENERAL INSTRUCTIONS :

1. All numbers must be written to the right side of the field, *i.e.*, ending on last column of field.
2. All alphabetic information must begin in first column of field.
3. Where a range is available (depth, time and size), if only one record is available write to the right of the dash (-).
4. Date in collection record must be Day, Month, Year, with no slashes or dashes, *ex.* 08121977.
5. Comments cards must be numbered 5 for collection records. No comments should go on line 4.

LOCALITY CODES (LOC) CARD 4

0 or blank - no data
1 - Queensland
2 - Lizard Island
3 - One Tree Island
4 - NSW (Including ACT)
5 - Sydney Harbour
6 - Victoria
7 - Tasmania
8 - South Australia
9 - Western Australia
10 - Northern Territory
11 - Lord Howe Island
12 - Norfolk Island
13 - Kermadec Island
14 - New Zealand
15 - Macquarie Island
16 - Antarctic
17 - New Guinea
18 - Solomon I., New Britain, New Ireland
19 - New Hebrides
20 - Fiji
21 - New Caledonia
22 - Indonesia
23 - 69 See other sheet

HABITAT CODES (H) CARD 4

blank - unspecified
F - Freshwater
E - Estuarine
M - Marine
C - Combined

BOTTOM CODE (B)

Card 4 (dominant)
blank - unspecified
C - coral
R - rock
G - gravel
S - sand
M - mud
V - vegetation
W - water
I - Invertebrate assoc.
O - Other (wood, etc.)

COLLECTION METHOD CODE (CR) CARD 4

blank - unspecified
RO - rotenone
EX - explosive
SN - seine
GN - gill net
BT - bottom trawl
MT - midwater trawl or plankton net
SE - selective, spear, quinaldine or hand net
HL - hook and line
FM - fish market
NL - night light
OT - other (beach washup, hand, etc.)
SH - shocker
TR - trap
CM - combined methods

FIELD DATA COMMENTS CARDS 5

Gear
Additional locality detail
Water temp. - ^{0}C
Salinity - ppt
O_2 - ml/l
Elevation in m
Tide
Wire out
Time at depth
Vegetation
Shore
Stream width
Any other

SOURCE CODES (S) CARD 4

blank - public donation
M - museum collection
E - exchange
F - state fisheries
U - professional
C - commonwealth
A - museum associates
P - purchased

SPECIES REMARKS CARD 7

subspecies
Author(s) of types
Date of description
Destroyed or exchanged details
Publication records
Other

SPECIES CARDS 6
TYPE CODE (T)

H - holotype
P - paratype
S - syntype
L - lectotype
R - paralectotype
N - neotype

SPECIMEN CODES (A)

blank - specimens
S - skeleton
C - cleared and stained
O - otoliths
K - skin
J - Jaws
A - specimen + cleared & stained
D - specimen + otoliths
E - jaws + skin
D - drum
T - tank
F - drum + otoliths
G - tank + otoliths
H - skeleton + otoliths
P - parts, egg cases, etc.

STATUS CODES (S)

E - exchanged
P - part exchanged
D - destroyed
L - lost

Figure 96: Fish Field Data Sheet (Hangay and Dingley, 1965)

Table 4: Morphometry and Life History Data Sheet (T. Laevastu, 1965)
(a) Species........................ (b) Date............... (c) Measured by...............

No.	Item					
1.	Locality	....	....	....	......	......
2.	Gear of capture	....	....	....	......	......
3.	Weight	....	....	....	......	......
4.	Total length	....	....	....	......	......
5.	Fork length	....	....	....	......	......
6.	Head length	....	....	....	......	......
7.	Snout to ins. 1st D.........................	....	....	....	......	......
8.	Snout to ins. 2nd D........................	....	....	....	......	......
9.	Snout to ins. anal	....	....	....	......	......
10.	Snout to ins. ventral	....	....	....	......	......
11.	Greatest body depth......................	....	....	....	......	......
12.	Diam. of eye.................................	....	....	....	......	......
13.	No. gill rakers	....	....	....	......	......
14.	Colour	....	....	....	......	......
15.	Stomach content: 15.1 Volume	....	....	....	......	......
	15.2 Composition	....	....	....	......	......
16.	Sex	....	....	....	......	......
17.	Degree of maturity	....	....	....	......	......
18.	Preserved: Scales	....	....	....	......	......
	Gonads	....	....	....	......	......
	Vertebrae	....	....	....	......	......
	Operculum	....	....	....	......	......
	Entire specimen.....	....	....	....	......	......
	Other (specify)	....	....	....	......	......
19.	Remarks:..					

//•••••••••//

Chapter 5
Terminology Used in Fish Identification

Common Terminology Used in the Fish Identification

Abdomen: Belly region or underside of the body between the breast and anus.

Abdominal: Located in the belly region or abdomen.

Accessory fins: Numbering 4-10, situated between dorsal and caudal fin.

Acute: One part adherent to another.

Adipose fin: Fleshy median fin without supporting skeleton like spines or rays (*Figure 97*).

Adnate: Sharp jointed.

Alae: Scale-like appendage at the base of caudal fin (*Figure 98*).

Anastomosing: Interjoining.

Anterior: Near to head, forward of another organ or opposite of posterior.

Antrorse: Facing forward.

Anus: Part of vent being the opening for excrement (*Figure 97*).

Apical: At the tip or apex.

Attenuated: Made thin, slender, reduced.

Auxillary scales: Small scales superimposed on larger ones.

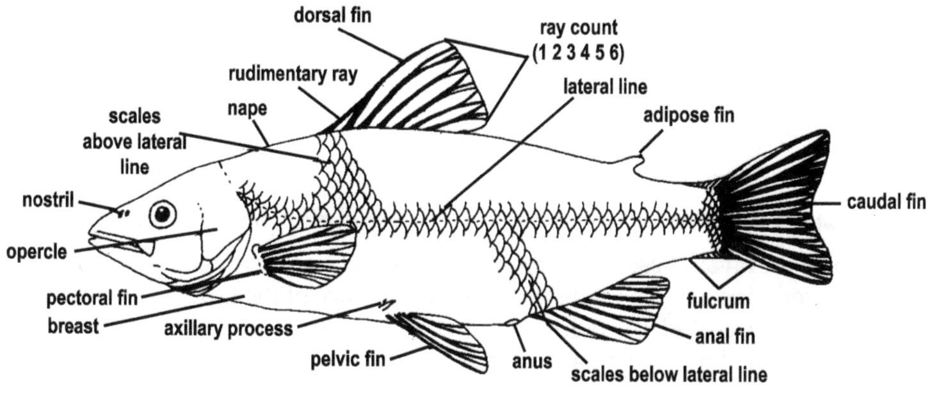

Figure 97: Generalised Fish Showing Some Important Features

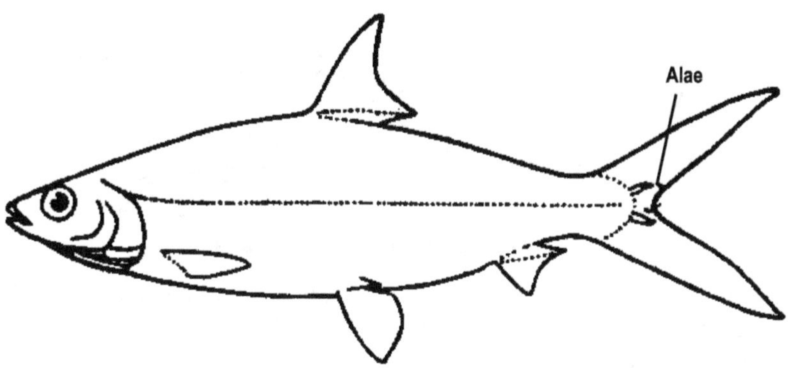

Figure 98: ***Chanos chanos***

Axilla: Cavity behind the base of fin.

Axillary lobe: Enlarged scale at the base of ventral fin (pelvic fin) (*Figure 99*).

Axillary process: Small process situated at the axil of the pelvic fins in some fishes (*Figure 97*).

Axis of body: An imaginary line running through the centre or mid of the body.

Barbles: Tactile filaments under chin or sides of mouth (*Figure 154*).

Belly: Region on underside of the fish between the pectoral fin and anus.

Bifurcated: Forked.

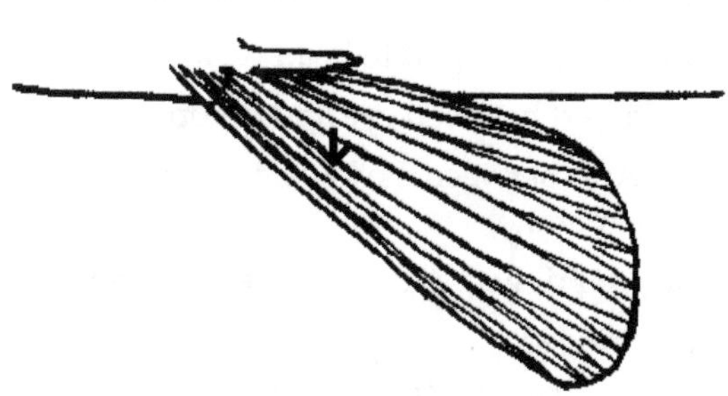

Figure 99: Axillary lobe

Bilobate: With two lobes.

Bony plates: Rhomboid plates arranged in rows on the body of few fishes (acipensaridae) (*Figure 12*).

Branchial: Pertaining to gills.

Branchiostegal: Pertaining to those anatomical parts which cover gills.

Branchiostegal membrane: Covers the gill opening during the swallowing of water.

Branchiostegal rays: These are numerous tiny thin bones arranged fan-wise from the lower edge of the opercle to the ventral surface of the head and covered by a thin skin called branchiostegal membrane. They serve to hold the branchiostegal membrane spread out (*Figure 100*).

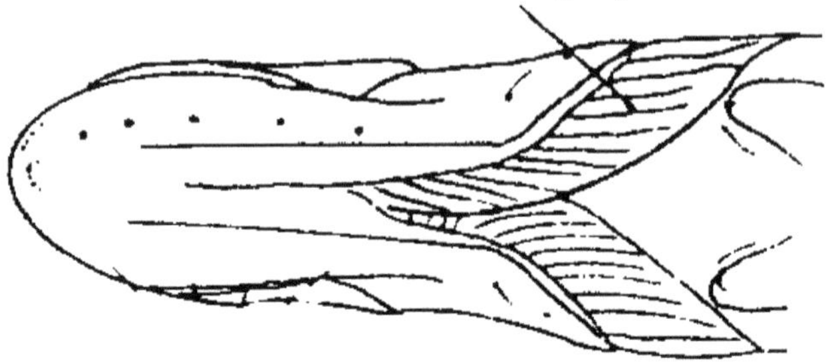

Figure 100: Branchiostegal Rays

Breast: It is the region which lies posterior to the isthmus and extends upto the base of pectoral fin. Heart lies in this region (*Figure 97*).

Buccal: Pertains to the mouth.

Cardiform: Coarse and sharp.

Carnate: Having keel or ridge along midline.

Caudal: Pertaining to tail.

Caudal fin: Median fin at the end of caudal part of body. It may be rounded, truncate, with a notch, deep incision, forked etc. (*Figure 97*).

Caudal peduncle: Part of the body behind anal fin (*Figure 112a).*

Cephalic: Concerned to head region.

Cheek: Part between eye and preopercle.

Chin: Area between the lower jaw and the point of attachment of the branchiostegal membrane.

Ciliated: Fringed with hair-like projections.

Cirri: Fringed.

Circumoral teeth: Horny teeth surrounding the mouth in lampreys.

Claspers: Modified ventral fins in male sharks and rays adapted for internal fertilisation.

Cloaca: A common posterior opening for anals and urino-genital openings.

Compressed: Flat from side to side.

Confluent: Joined together.

Cranial: Pertaining to skull.

Crenulate: With edge slightly grooved.

Cristae: With a crest.

Ctenoid scales: Bony scales with spines.

Cultrate: With a knife-edge.

Cuneate: Wedge-shaped.

Cutaneous: Pertaining to skin.

Cycloid scales: Bony scale with a smooth surface and smooth edge, without spines.

Dendritic: Tree-like.

Dentate, denticulate, dentigerous: With tooth-like projections.

Depressed: Flat from top to bottom.

Depth: The vertical distance through part of the body measured at right angles to the long axis of the fish *e.g.*, head depth, body depth.

Dermal: Pertains to skin.

Distal: Remote from point of attachment.

Dorsal: Pertaining to the back and opposite of underpart or ventral.

Edendate: Without teeth.

Falcate, Falciform: Scythe-shaped or sickle-shaped, with concave outer margin.

Filiform: Thread-like.

Fimbricate: Fringed at the margins.

Fin rays: Fin supports, may be branched or unbranched.

Fins: Median (dorsal-D, anal-A, caudal-C and adipose) or paired (P-pectoral and V-ventral).

Fontanel: An opening.

Fork: A place where dorsal and ventral lobes of caudal fin separate.

Fossa: A pit or depression.

Frenum: Fleshy bridge across the groove between the premaxillaries and snout (*Figure 101*).

Fulcrum: Enlarged base of the caudal (*Figure 97*).

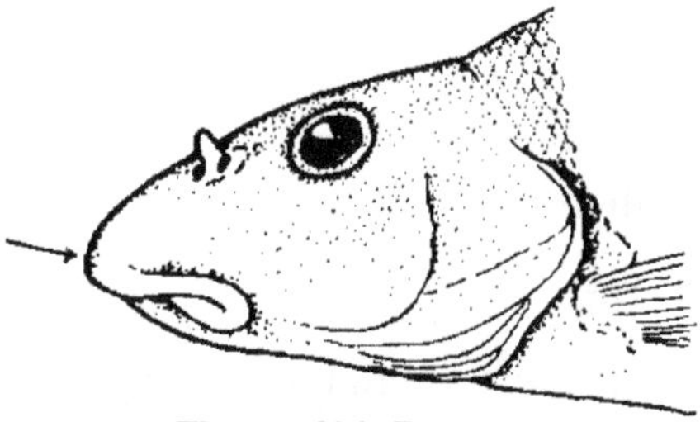

Figure 101: Frenum
(After Eddy and Underhill, 1969)

Fusiform: Spindle-shaped.

Ganoid scales: These are rhomboid or diamond-shaped.

Genipores: Very small pores on the head of Gobiidae and some other fishes.

Gill: Filamentous respiratory structure on each side of the pharynx or throat of a fish.

Gill cover: Covers the branchial cavity and as a rule of four bones–operculum, suboperculum, preoperculum and interoperculum.

Gill opening: A slit of various sizes through which water issues from gill cavity to the outside when it is propelled through gills during respiration.

Gonopodium: Modified fin ray for transfer of spermatophores to the body of female.

Gular plate: A plate covering the upper part of throat.

Hair-like teeth: Teeth resembling the hair.

Horny: Having nature of horn or fingernail.

Horny covering: A pad of cartilagenous cover found on inside lips of some fishes.

Hyaline: Colourless, clear.

Hyoid: Series of bones anterior to the gill arches and extending forward in the floor of the mouth.

Inferior: Beneath or somewhat behind another structure.

Interbranchial space: Part dividing the gill cavities into right and left.

Inter-orbital: Space above and between eyes.

Isthmus: It is the interspace which separates the gill openings of the left and right side and which lies anterior to the breast.

Jugular: Pertaining to throat.

Labial: Pertaining to lips.

Lanceolate: Spear-shaped.

Lateral: Towards the sides.

Lateral line: A sense organ, seen as a row of opening in the scales, skin or plates. The openings are arranged in one row, several rows, with or without interruption.

Lower jaw: Consists of three bones–dentale, articulare and angulare.

Median longitudinal groove or fontanel: These are two longitudinal externally visible long depression on the centre of the head and extend from the snout to the base of occipital process (*Figure 102*).

Molar/Molariform teeth: Blunt and rounded grinding teeth.

Multiserial: Arranged in many series.

Nape: It is back of neck (*Figure 97*).

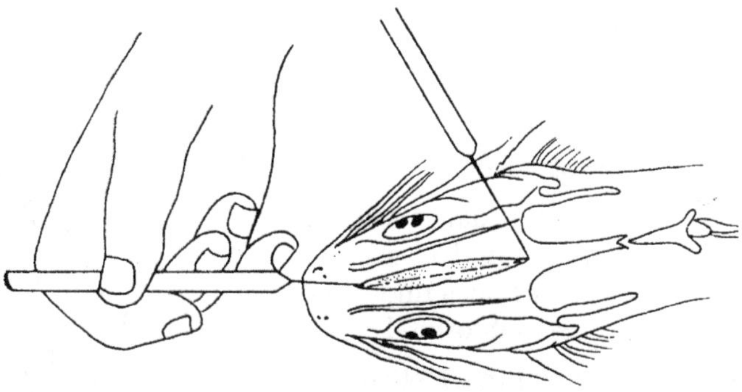

Figure 102: Median Longitudinal Groove (Extent of Fontanel)

Nasal: Pertaining to nostrils (*Figure 97*).

Non-protractile premaxillary: Premaxillary not completely separated from the snout, connected by a fleshy bridge or frenum *(Figure 103)*.

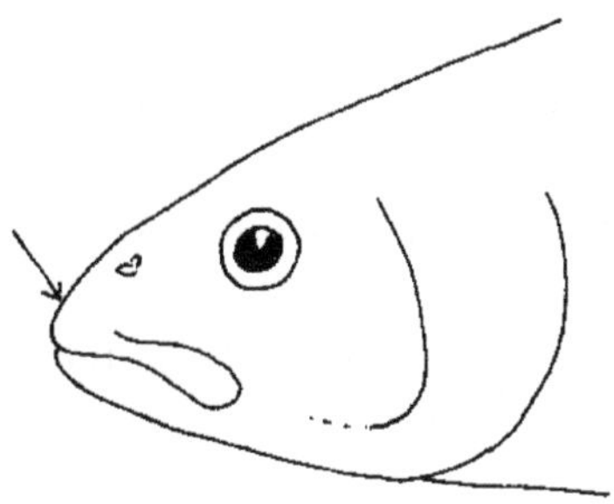

Figure 103: Non-protractile Premaxillary

Nuchal: Pertaining to the nape.

Occiput: It is the point on the mid-dorsal line which joins the head with trunk.

Ocular: Pertaining to the eyes.

Opercular bone: Membrane bones making up the gill cover and consists of opercle, preopercle, subopercle and interopercle (*Figure 104*).

Orbit: Bony eye socket.

Osseous: Bony.

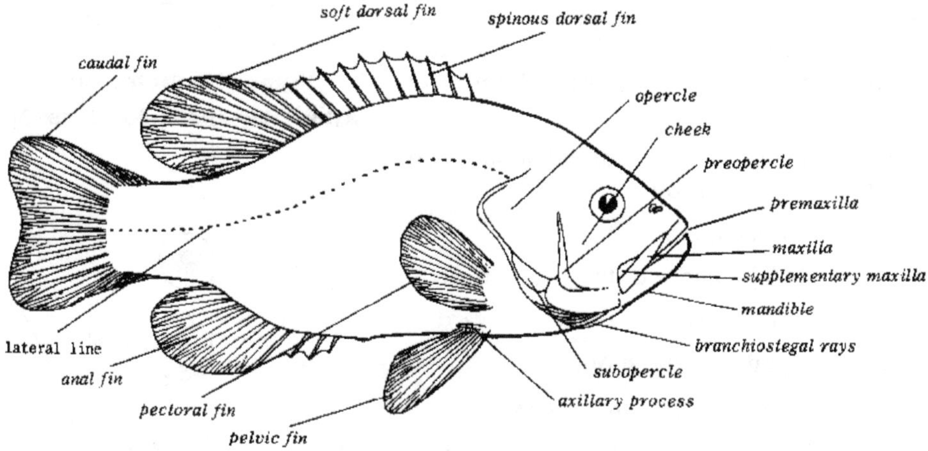

Figure 104: Opercular Bones

Palate: Roof of mouth.

Papillae: A small fleshy projection.

Pectinate: Shaped like a comb.

Peduncle: Basal support of the tail.

Periorbital bones: Ossicles situated around eye; anterior, the preorbital is usually enlarged and the smaller ones under the eye are generally called sub-orbital.

Pharyngeal teeth: These are throat teeth borne on the pharyngeal bones in cyprinids and cichlids which are modified gill arches. Each arch is with 1-3 rows of teeth (*Figure 105*).

Photophores: Luminiscent organ in deep water fish.

Plicate: Covered with creases or folds (plicae) *(Figure 106).*

Predorsal scale: Scale on the back before the dorsal fin.

Proboscis: Elongated part of the mouth projecting forward.

Procumbent: Projecting forward.

Procurrent: Leaning forward.

Protractile: Upper jaw without frenum, with a groove present between snout and upper lip *(Figure 107).*

Protrusible mouth: Occurs in fish in which premaxillary and maxillary bones are oftenly free and capable of moving forward (*Figure 108*).

Pseudobranch: Rudiments of gill situated on the inside of gill cover.

Pyriform: Pear-shaped.

Ray, Radial: A jointed rod that supports fin.

Retrorse: Curled downward and backward.

Rostral: Pertaining to the snout.

Rostrum: A projecting snout or beak.

Rudimentary: Small, tiny, incomplete or malformed.

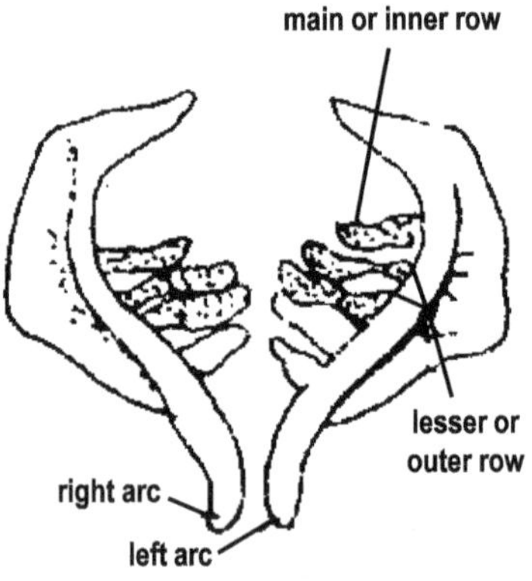

Figure 105: Pharyngeal Teeth

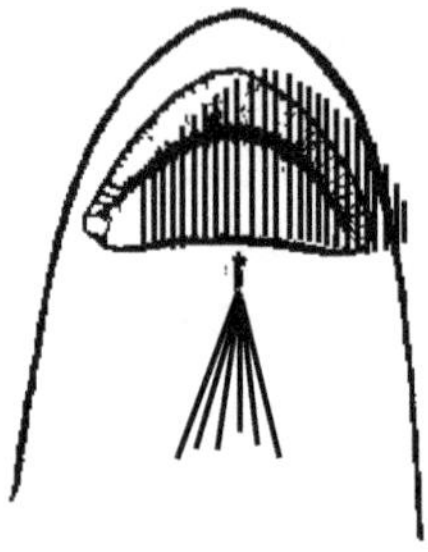

Figure 106: Plicate

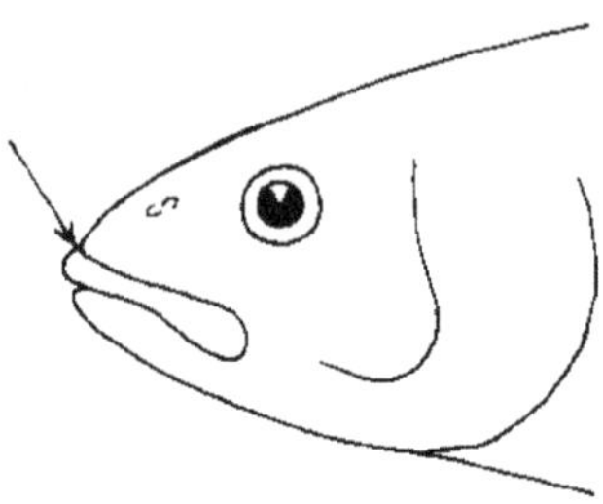

Figure 107: Protractile (After Eddy and Underhill, 1969)

Figure 108: Jaw Bones of *Spicara maena*

Rudementary soft rays: Soft rays at the front of the fin that are usually short and not divided at their tips *(Figure 97).*

Scute: An external horny or bony plate.

Serrae: Notches or teeth like a saw (pectoral spine) *(Figure 109).*

Snout: Whole of the head infront of eyes.

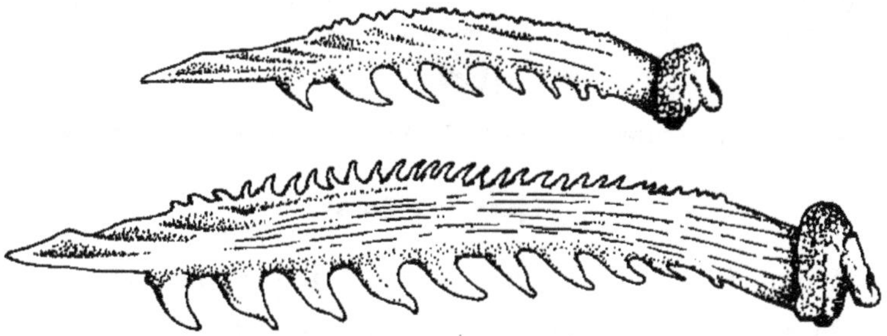

Figure 109: Serrae

Spine: A fin ray which is ossified, stiff and pointed.

Spiracle: An opening situated behind the eye in sharks and rays that leads into oral cavity.

Sub-orbital process: Processes of sub-orbital bones directed inward and supporting the eyeball.

Sub-terminal: Located below and some distance behind the end of a structure (*Figure 110*).

Symphysis: Site of union of left and right lower jaw bones.

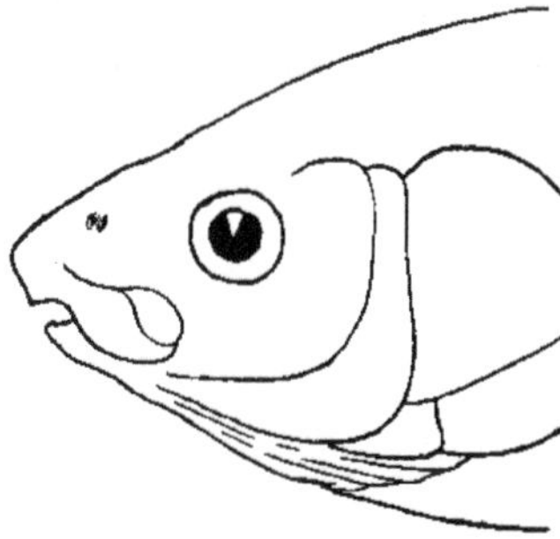

Figure 110: Sub-terminal

Tail: Part of the fish's body from vent posteriorly and does not includes caudal fin.

Tautonymy: Same name for both genus and species *e.g., Pangasius pangasius.*

Terminal: Located at the end of a structure.

Thoracic: Pertaining to chest.

Throat: Area on ventral side of the body between the base of pectoral fins and the point of attachment of branchiostegal membranes.

Tricuspid: With three points.

Trunk: Body region from the posterior margin of the sperculum to the anus (*Figure 111*).

Tubiform: In the form of tube.

Undulated: Wavy.

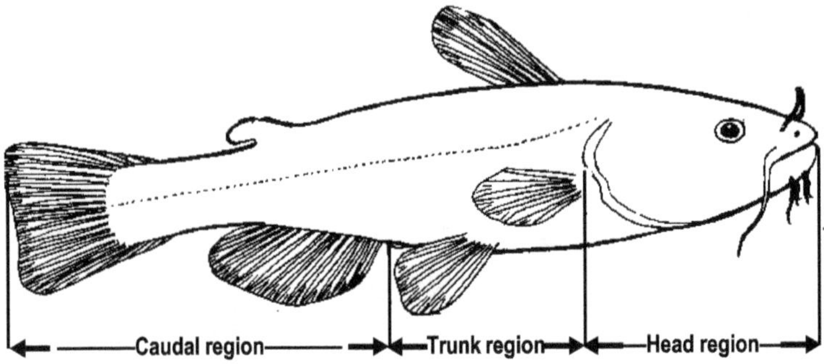

Figure 111: Trunk

Vent: An opening, in which intestine terminator, it is situated in front of anal fin.

Vermiculating: Forming a pattern of wavy worm like lines.

Vestigial: Rudimentary.

Villi: Minute finger-like processes.

Viviparous: Producing living young ones from within body.

//•••••••••//

Chapter 6
Measurements (Morphometric, Meristic and Biometric) and Counts

Measurements

A wide array of morphometric, meristic and biometric features that have been in use and considered very important for understanding a fish include:

Morphometric [Figure 112(a–f)]

Anal length: It is distance from the anteriormost part of the body to the anterior limit of anal aperture.

Anal fin length: It is the length of the ray from the anterior insertions of the anal fin.

Anal fin base: It is the distance between the anterior and posterior insertions of the anal fin.

Body depth: It is the measurement from a point in the body of the fish on its back where its height is greatest to a straight line to the ventral surface or profile. The fleshy or scaly structures pertaining to the fin base are excluded. It need not necessarily be in the middle of the fish.

Caudal fin height: It is the distance between the stretched lobe of the caudal fin.

Caudal peduncle scale count: It denotes number of scale rows crossing the circumference line of the body at the level of narrowest part of caudal peduncle.

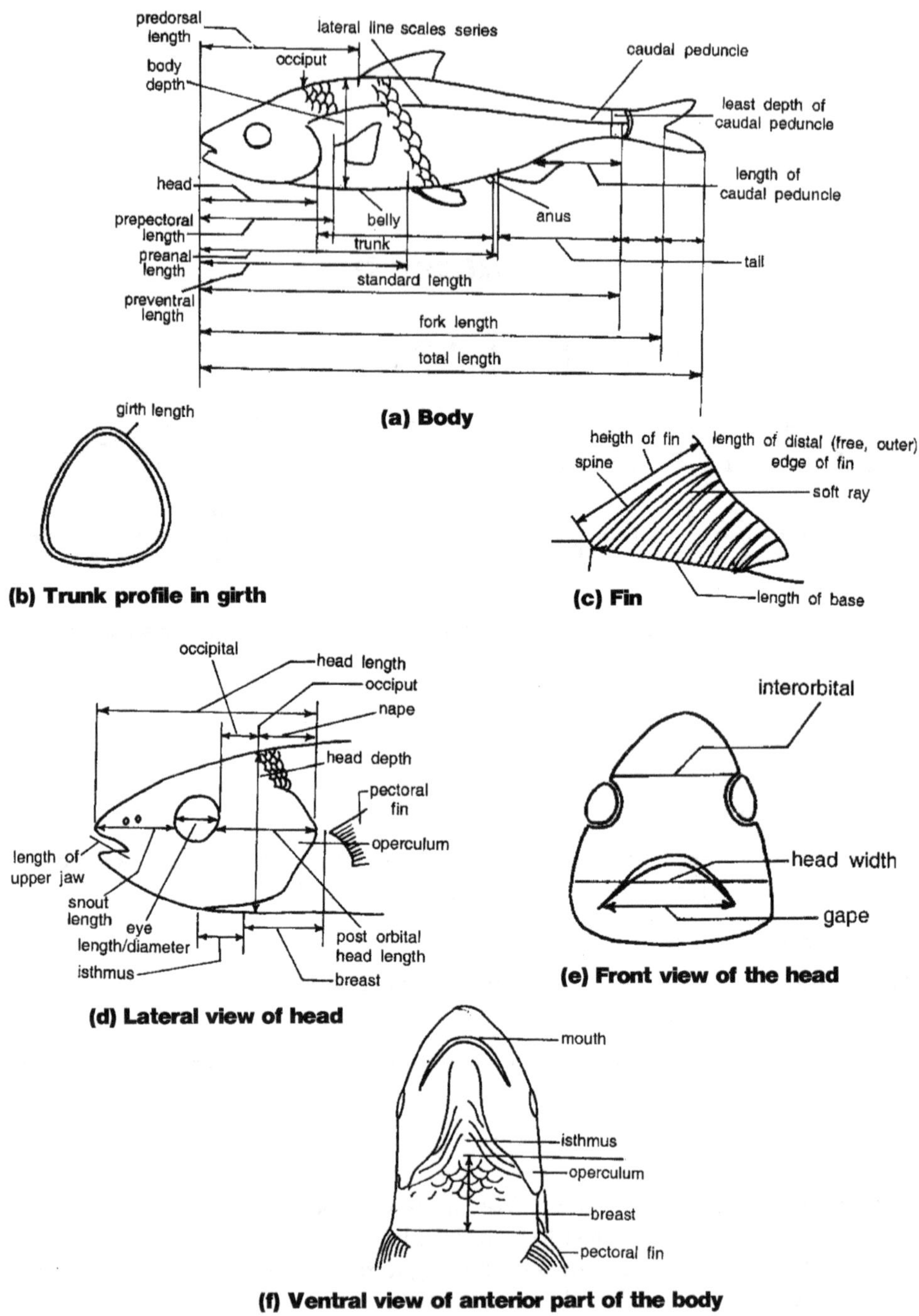

Figure 112(a–f): Figures Depict All Measurements Related to Length and Height

Depth at dorsal fin: It is the depth of the body perpendicularly through the anterior insertion of dorsal fin.

Depth at pectoral fins: It is the depth of the body perpendicularly through the base of the pectoral fin.

Depth of anus: It is the depth of the body perpendicularly through the anus.

Depth of mouth: It is the distance between the dorsal and ventral surfaces of the head, perpendicularly through the posterior angle of the mouth.

Depth at eye: It is the distance between the dorsal and ventral surfaces of the head, perpendicularly through the eyes.

Dorsal fin height: It is the length of the largest fin ray or spine of the dorsal fin.

Dorsal fin base: It is the distance between the anterior and posterior end of dorsal fin taken along the base of the fin.

Eye diameter: It is the distance between front and rear margin of the eye in the longitudinal axis.

Fork length: Distance between anteriormost part of the body and the point of bifurcation of forked caudal fin.

Girth length: It is the distance covering the circumference of the body at its deepest level. This measurement sometimes indicates maturity of the species particularly in female fishes.

Gape width: Also termed as width of mouth. The greatest distance of the mouth opening when the mouth is closed.

Height of dorsal or length of dorsal fin spine: It is measured from anteriormost point of junction with the body to anterior tip of fin or spine even if other rays do not reach this point.

Head width: Greatest distance across the head with gill covers held in normal position.

Head depth: It is the perpendicular distance from the end of the nape to the ventral side of the head.

Head length: It is the distance in a straight line between anteriormost part of snout or upper lip (whichever is extended farthest forward) and the posteriormost edge of the opercular bone.

Interbranchial space: Part dividing the gill cavities into left and right.

Inter-orbital width: Least distance between the bony rims on the inner margins of the eye.

Length of pectoral and pelvic fins: It is measured between its origin or place of insertion into the body to its extreme tip.

Length of longest fin ray: It is measured from the structural base of longest fin ray to tip.

Length of caudal peduncle: It is the distance between rear point of anal fin base and the base of the caudal fin.

Length of base of spinous or soft dorsal, adipose dorsal, anal fin: The greatest distance measured in a straight line between the anteriormost and posteriormost points of junction with the body.

Least depth of caudal peduncle: It is the shortest distance of height of caudal peduncle.

Lower jaw length: It is the greatest length of the lower jaw.

Length of pectoral and pelvic fins or spines: Distance of their origin or point of insertion into body to the extreme tip of fins/spines.

Occipital: It is the distance between the two prependicular lines touching the posterior end of the orbit and the occiput.

Pre-pelvic distance: A straight line distance from the mid-point of the base of pelvic fin to the anteriormost point of head.

Pectoral fin length: It is the length of the ray from the dorsal base of the pectoral fin.

Pectoral fin base: It is the distance between the dorsal and ventral insertions of the pectoral fin.

Pre-orbital or snout length: It is the distance in a straight line between anteriormost part of snout/upperlip and the anterior margin of the orbit.

Post-orbital length: It is the distance between posterior margin of the orbit and the posteriormost edge of opercular bone.

Pre-dorsal length: It is a straight line measurement from the tip of snout/upper lip to the structural base of anteriormost first dorsal fin ray.

Post-dorsal length: It is a straight line distance between the structural base of dorsal fin to the flexure line of the body over the caudal peduncle or end of the vertebral column.

Pre-pectoral length: It is the distance between the anteriormost end of the body and the front point of the pectoral fin base.

Pre-anal length: Distance between anteriormost end of the body and the front point of anal fin base.

Standard length: Distance between anteriormost part of the body (snout/premaxilla/upper lip) and the base of caudal fin or end of vertebral column.

Sub-orbital width: Least distance from the lowermost margin of the orbit to sub-orbital or pre-orbital margin.

Total length: It represents maximum elongation of body from end to end. It is the distance between tip of the snout or upper lip and the posteriormost extremity of the body (tip of the longest caudal lobe if it is forked and has unequal lobes). The measurement is a straight line and is not taken over the curves of the body.

Trunk circumference scale count: It denotes the number of scale rows that will cross the circumference line of the body at the level of dorsal fin.

Upper jaw length: It is the greatest length of upper jaw from the anteriormost point of premaxillary to posterior point of maxillary.

Ventral fin length: It is the distance from the anterior insertion of the ventral fins.

Ventral fin base: It is the distance between the outer side bases of the ventral fins.

Width of premaxillary teeth band: Maximum width of teeth band.

Meristic

Among the meristic studies, osteology (study of bones) has been an accepted instrument for identification. Bones prepared from integral part of vertebrate, animal collection and assembled skeletons are most beneficial for study or display purpose. The preparation of skeleton, done by any of the following methods, is a complete, unmutilated manual. For the skeletal preparation, fish specimen is preserved whole or kept frozen. Fish is then eviscerated and its flesh removed carefully from both the sides of body. Care should be taken to prevent the damage of fine bones. Since most fish dry easily and quickly, it is often more practical to leave a considerable amount of flesh attached to the bones rather than damage the skeleton by vigorous carving and scraping.

Maceration

Cold water maceration: It can be done in case of large specimen when the specimen is kept completely submerged in water until all the flesh, ligaments and so on rot away. However, the disadvantage of this method is unpleasant smell and it can be reduced by adding agar agar which absorbs most of the unpleasant odours.

Warm water maceration: In this case the vessel is equipped with a thermostat regulated heating element which keeps the water temperature at 37^{0}C. This process is quick. A grill is fitted approximately 30-40 mm above the bottom of the vessel to keep the bones away from the sediments which will collect below it.

Maceration by cooking: It is the quickest method of removing the flesh. A large aluminium or stainless steel pot with a grill insert is suitable. The container should be filled with water, 150 gm sodium carbonate added for each 5 litres and the bones boiled for 3 hours. The time also varies according to the size of the specimen; the larger it is, the longer it takes to cook. When the flesh starts falling away from the bones, the specimen are cooked long enough so that every bit of flesh is removed. After cleaning, specimen is rinsed and dried.

Dry maceration: In this method, the bones are buried in beach sand in a brick enclosures. Individual skeletons are placed in flower pots and buried. Maceration takes 10 months, and after 18 months hair and ligaments are also macerated.

Chemical Maceration

Chemicals are also used in maceration. **Ammonium** and **sodium hypochlorite** are used in maceration of skull of young animals. **Sodium perborate** with 60–70 gm/l for smaller specimen and 70–100 gm/l for larger specimen is used in boiling water for maceration.

Potassium hydroxide: Fleshed bones are soaked in 5 per cent solution of KOH and water at 35⁰C–40⁰C for maceration.

Sodium hydroxide: The specimen which are previously stored in carbolic acid are immersed for 45 minutes in 1 per cent NaOH-water solution at 75⁰C.

Tryspin: Two-heaped teaspoons of tryspin and one-heaped teaspoon of sodium sulphide in 4 litres of water at 35–40⁰C for 2–3 days helps in maceration of bones.

Papain: This is used to macerate smaller specimens. The specimen are kept in water at 37–45⁰C and small amount of papain is sprinkled on the surface. The vessel is covered with close fitting lid. The maceration of small skeleton takes 2–3 days.

Degreasing

After maceration, the bones can be degreased, bleached (if necessary) and dried. The degreasing can be done with or without an instrument. Trichloroethylene is a widely used degreasing agent. Genklene (1.1.1. trichloroethane) is commercially used. Sodium bicarbonate is the least dangerous substance suitable for degreasing.

Bleaching

The appearance of skeletal specimens is enhanced by bleaching. Although it is not desirable to bleach the bones completely white, but even creamy white colour is advantageous, not only from an aesthetic point of view, but also because a relatively uniform tone renders the specimen more practical for the study purpose. The bleaching can be done with hydrogen peroxide.

Drying

After bleaching, the specimen should be thoroughly rinsed and dried. Rapid drying should be prevented as it can cause the cracking of bones.

Staining Skeletal Material

The real advantage of clearing and staining of skeleton is to observe very small and delicate parts which would be difficult to see and study otherwise when unstained.

The bones are stained red with alizarin red S, which selectively stains the calcium. In this method, specimens are fixed in 80 per cent alcohol for a minimum of a week after which they are skinned, all scales removed and fish eviscerated.

After fixation, the specimen is washed for an hour and placed in 1–2 per cent solution of KOH for 2–5 days. During the time the tissue turns brown and any fat present will stand out as white patches and can be removed using forceps and scalpel. Alternatively, fat can be removed by placing specimen in acetone for few days after fixation and before placement in KOH. The specimen becomes soft and jelly like and any movement may cause it to disintegrate.

The staining solution is prepared by adding 10 gm of KOH to alizarin red S (0.1 gm) in 1 litre of distilled water. The volume of the solution should be 10 times the mass of specimen. The specimen should be kept in stain for 1–2 days after which the solution should be replicated with fresh 1 per cent solution of KOH. The tissue as well as bones will become red in colour, immersion in KOH will destain the tissue. Immersion in several solutions of KOH may be necessary before only the bones appear red. When this has been achieved, 50 per cent of the KOH can be removed and equal volume of glycerine added. The specimen may float, so the jar should continue to be inverted until the specimen remains at the bottom of the containers. After one day in this mixture, remove 50 per cent of the solution and replace it with same volume of fresh glycerine, repeating this procedure 3 to 4 times before adding 100 per cent glycerine. To prevent the growth of mould, add a crystal or two of thymol. This technique is very cheap and easy to perform.

Enzyme Method

Sometimes fish do not clear and stain well using the KOH method, so specimen may be cleared using an enzyme called tryspin instead of KOH. Specimen should be injected 10 per cent neutral formalin, then left to soak for 12–24 hours. The animal then cleansed of scales and eviscerated and is soaked in acetone to remove fat deposits. The specimen is then placed into a tryspin solution (add 125 ml saturated sodium borate into 375 ml distilled water and 1 gm tryspin powder). Leave the specimen in this solution until the tissue begins to clear supernatent liquid from the stock. Alizarin red S can now be added and the specimen is left till three quarters of tissues are transparent. The specimen is then washed in water and placed in a bath containing 1 per cent solution of KOH which will cause the colour to appear in the bones and tissue. When KOH is stained, replace with fresh KOH until the de-staining is complete. Bathing in a mixture of KOH and glycerine can now take place until final storage in 100 per cent glycerine.

Biometric Index

The number of time each body parameter goes into the total length of the fish is taken as the biometric index. For each body parameter/character in relation to total length, a mean biometric index is to be calculated. If, in all the size groups, a particular character/parameter shows constantly increasing ratios, the character in question exhibits positive allometry and the reverse condition is known as negative allometry.

A constant index in any of the biometric indices indicates that the growth of the biometric character in relation to its reference length is isometric.

To study the morphometric and meristic characters, the regression method is employed with formula: $Y = a + bX$,

where,

Y: Variable character (such as total length, head length etc.)

a: Constant value to be determined

b: Regression coefficient

X: Total length

The values of a and b are determined by following formulae:

$$a = \bar{y} - b\bar{x}$$

$$b = \frac{\sum xy - n\bar{x}\bar{y}}{\sum x^2 - n(\bar{x})^2}$$

where,

n: Total number of length groups

x: Mean of X

y: Mean of Y

Counts

Meristic Counts

The meristic characters are those which are countable *e.g.*, vertebrae, scales, fin rays etc. and retain their species specificity. The fin ray and scale count very precisely constitute the fin formulae and are dealt in detail as under:

Fin rays: Fin rays are of two types–undivided rays or spines and divided rays or soft rays also called as "rays". It is very easy to detect the ray and spine in large specimens but in smaller specimen, microscope and/or hand-lens is required.

To denote various fins and other structures, the standard abbreviations used are as under:

A: Anal fins.

B: Branchiostegal rays (curve bony rods found on either side of ventral aspect of the head).

C: Caudal fin.

D: Dorsal fin.

Ll: Lateral line scale.

Ltr: Lateral line transverse scales.

Lr: Lateral line interrupted or found in scaleless fish.

P: Pectoral fin.

V: Ventral pelvic fin.

Fin Formula

An oblique stroke is used to separate spine and the rays. Spine counts are expressed by Roman numerals (I–X), while soft rays are expressed by Arabic numerals (1–10). Spines are the rays that are unbranched and unsegmented structures. Soft rays are the rays that are often branched and always biserial (two lateral components paired on the midline), segmented and flexible. In case where both spines and rays coexist, Roman and Arabic numerals are used to give their counts in the same order. Often capital and small Roman numerals are used to distinguish between strong and weak spines.

Counting fin rays in case of rayed dorsal fin, the posteriormost ray is taken as the first ray and the count is done towards the anterior direction. Even the smallest ray at the anterior end of fin base is counted.

In case of anal fin the direction is from anterior to posterior.

A few examples can be cited to explain this:

D II 18:	Means that dorsal fin has 2 strong spines and 18 soft rays.
D ii 18:	Means that dorsal fin has 2 weak spines and 18 soft rays.
D 10-12:	Only soft rays are present and their number varies from 10–12.
D iii-iv 9-11:	Means that number of weak spine varies from three to four and that of soft ray from nine to eleven.
D_1 I4:	Means first dorsal fin with one strong spine and four soft rays.
D_2 5:	Means second dorsal fin with five soft rays only.
D ii 15/0:	Means first dorsal fin has two weak spines and fifteen soft rays but the second dorsal fin is an adipose fin without rays.
Dx+1 26-27:	Dorsal fin has 10 weak spines and a soft part while the second dorsal fin has 1 strong spine and 26 or 27 soft rays.
A + C 100:	Means anal fin is confluent with caudal fin and total count of 100 is soft rays.

Lateral Line Scale Count

Scale counts with reference to lateral line are also used in taxonomy. Number of scales along lateral line and also in transverse row above and below it on each side is expressed by *scale formula*.

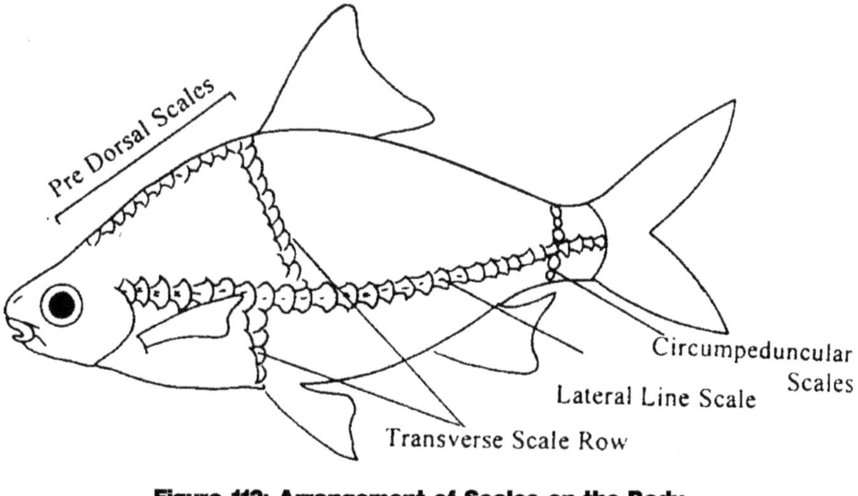

Figure 113: Arrangement of Scales on the Body

Lateral line scales are counted as the number of perforated scales along the lateral line. This lateral line when found interrupted can be written as L.r. and this symbol is also used in scaleless fishes (*Figure 113*).

The lateral line transverse scales (L.tr.) can be counted as follows:

The scales above lateral line can be counted from the origin of dorsal fin and counted downward and backward to the lateral line scale. The scales below the lateral line are counted upward and forward from the origin of anal fin *e.g.*, Ltr. 14– can be written as 8/1/5 or 8½ 5½.

Example can be: If we write

80 5/7 : It means 80 scales in lateral line, 5 scales above it and 7 scales below it.

Ll 30: Means that there are 30 scales on lateral line.

Ltr 16: Means that the number of lateral line transverse scales falling on a line from the origin of the dorsal fin to the origin of ventral fin is 16.

Ltr 6/1/4: Means that 6 scales are above lateral line scale and 4 scales are below it.

//••••••••••//

Chapter 7
Morphological Variations

Fish identification mainly depends upon the morphological characters that include the external features and their variations. It is because of these characters that species can be differentiated from one another and ranked at appropriate places in the taxonomic hierarchy.

These variations make us understand differences at the lowest level of identification as well as incidence of their occurrence in community and habitat. These variations may occur in integument/body form/head region/body region/ trunk and/tail.

Integument

Skin, an outermost covering of the body is the first line of defence against disease and other external stresses. It affords protection against environment and has respiratory, excretory and osmoregulatory functions. Housed in skin are the devices of colour, electric organs, venomous glands, phosphorescent organs and mucous glands.

In most of the fishes, skin is covered with an exoskeleton of scales. These scales find much use in taxonomy. Their types, distribution, modification or absence as well as scale count have diagnostic value in it.

Structure

The skin of a fish for long was considered composed of two layers, but recent observations suggest it to be made of three distinct layers, an outer cuticle, followed by epidermis and innermost dermis or corium (*Figure 114*). The cuticle is a delicate, extracellular layer. This layer comprises of cell protoplasm, sloughed malphigian cells and some secretion of mucus (*Figure 115*). The thickness of cuticle varies

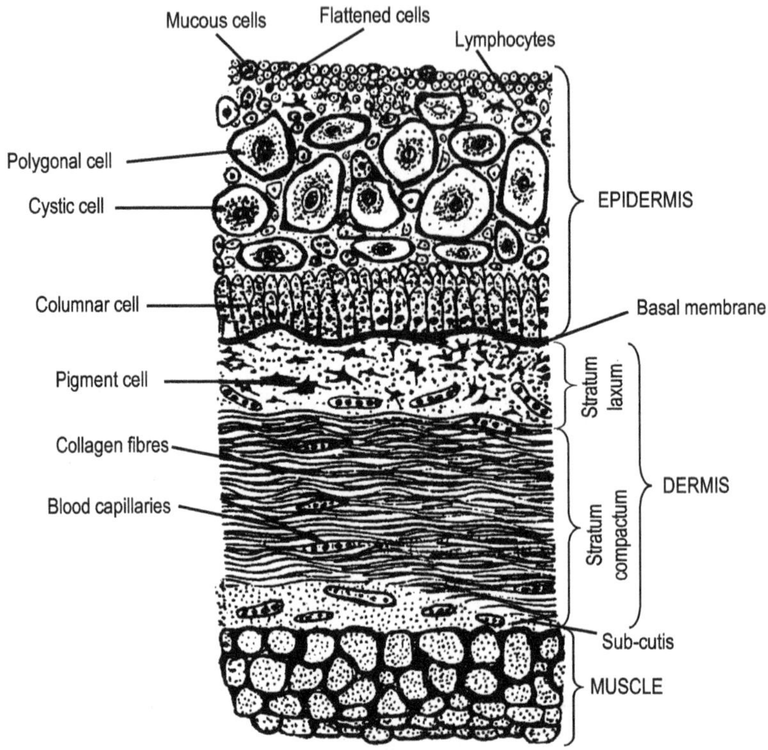

Figure 114: ***V.S.*** **of Fish Skin (Generalised)**

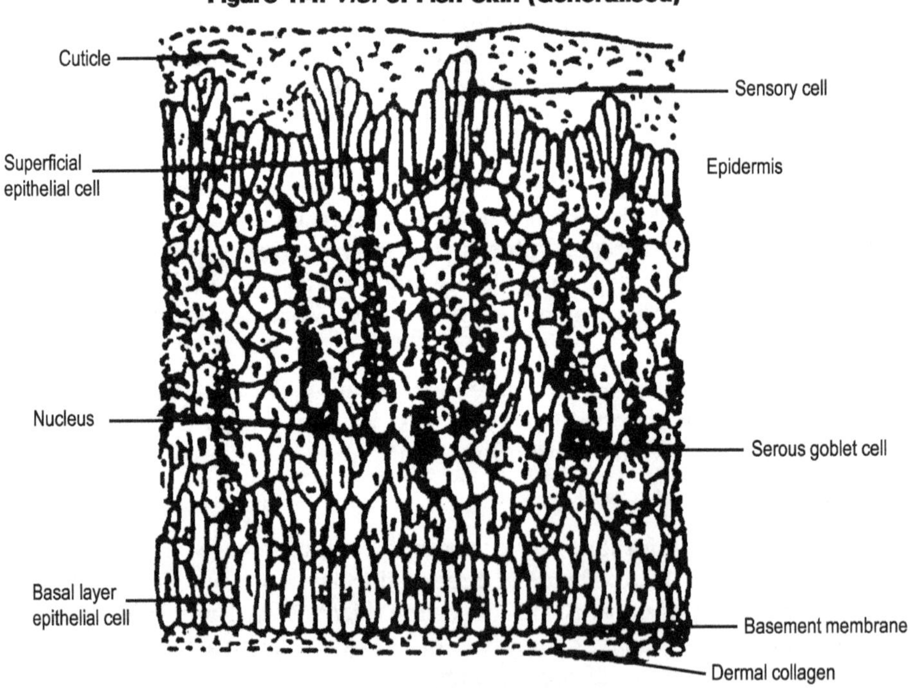

Figure 115: ***Trigla*** **sp. Skin Dermis**
(After Kapoor and Khanna, 2004)

considerably. The epidermis is ectodermal in origin and is similar in many ways to the lining of the human mouth. Typically it is composed of several layers of flattened cells (stratified epithelium) of which the deepest layers are made up of columnar cells forming **stratum germinativum** and is the zone of active cell growth and multiplication. This layer is fertile and here cell multiplication goes on all the time to replace from within the outermost scalised layer of cells as soon as these are worn-off, and to provide for growth. A superficial layer of dead horny cells, forming the **stratum corneum** in terrestrial vertebrates, is not present in fishes. This layer encompasses a diversity of other cells like goblet cells, sac cells, club cells, granular cells, thread cells, venom cells, ionocytes, markel cells, epithelial sensory cells, cement secreting cells and intrusive cells.

The thickness of epidermis varies with species, age or position on body. In most fish, especially benthic species, the ventral epidermis is thick. It is usually thick in naked species or species which have vestigial scales. In scaly fish it covers and follows the outline of scales. The epidermis over fins and tail is qualitatively different from other body and lacks more specialised cells.

The dermis is mesodermal in origin and contains blood vessels, nerves and cutaneous sense organs and connective tissue. It consists of two layers. The upper layer is made up of loose connective tissue and is known as **stratum spongiosum** and the lower layer which is thick and dense is known as **stratum compactum**. Increased production of mucus during reproductive period serves to protect the body regions injured during abrasive behaviour (*Pimephales promelas*). Numerous tubular or flask-shaped cells are scattered among the epidermal cells and may even extend into the dermis. These cells secrete mucin, a glycoprotein, which mixes with water to form a thick, slimy mucous covering the whole body. Mucous cells develop from the **stratum germinativum** and migrate to the surface. Their number, size and kind varies in different species of fishes. The fishes having no scales or poorly developed scales possess large number of mucous cells. The integument is attached to the underlying musculature by a layer of aveolar tissue, the hypodermis, which provides a degree of flexibility allowing lateral movement between skin and underlying tissue.

The colouration in fishes is due to the presence of skin pigments. Two types of cells, chromatophores and iridocytes are present in the dermis, which contain pigment granules. Because the fishes lead an aquatic life the epidermis is generally soft and cornification is a rare case. A peculiar pearl organ appears in the skin (minnows–cyprinidae) in the breeding season. In some teleosts (*e.g. Trigla* sp.), the epidermis contains chemoreceptor cells which resemble the taste buds. The excretory chloride cells occur in the epidermis of *Trigla* sp. In elasmobranchs, pterygopodial glands are present in the claspers of males.

Functions

The mucus covering the body of the fish performs several important functions:

- It lubricates the fish so as to reduce body friction in water, while swimming, thus enabling the fish to move with greater speed.

- It protects the body from parasites, fungus, bacteria and other micro-organisms.
- In turbid waters, the mucus has the power to precipitate the particles in suspension especially in contact with the gills and thus prevents the fish from suffocation and helps in its survival.
- The skin and mucus secreted by its glands help the fish in regulating the exchange of water and ions between the body fluids and the surrounding medium.
- Discrete glandular patches (multicellular glands comprising of goblet cells) situated under the operculum of labrids (*Labroides phthirophagus*) and scarids (*Scarus dubius*) produce transparent mucoid shirt or cocoon that fish wears at night. This envelope protects against predators at night.
- Large mucous goblet cells in the tip of tail of semi-sendentary garden eel (heterocongridae) produce mucous to line their burrow.
- The African lung fish (*Protopterus* sp.) prepares the hard cocoon with the mucus secreted by its skin for its summer sleep.
- Certain species (*Gasterosteus*, *Betta* and *Macropodus*) make use of mucus to prepare nest for egg laying.
- The mucus repairs the surface wound as it contains large number of lymphocytes.
- The skin also functions as an important respiratory organ in certain species like *Anguilla* and climbing perch.
- Glandular cells of epidermis are modified to form poison glands and are used for offence as well as defence (*e.g.*, scorpion fish and toad fish).
- Light producing phosphorescent organs are also derivatives of the epidermis.
- Chromatophores present in the dermis give beautiful colour pattern to the body.
- Exoskeleton in the form of scales, plates or denticles are important derivatives of skin and provide protection to the body.
- The skin also absorbs dissolved nutritive substances from the surrounding water.

Scales

Based on the presence/absence of scales the body of fish may be naked/ intermediate (scale distribution restricted to limited portion of body)/scaly.

Naked

Some fishes are 'naked'. In the technical sense they have no scales on their body as in lampreys (petromyzontidae) (*Figure* 3) and freshwater catfishes (ictaluridae) (*Figure 116*), and even among some sticklebacks (*Pungitis* sp.) the body is naked (*Figure 117*).

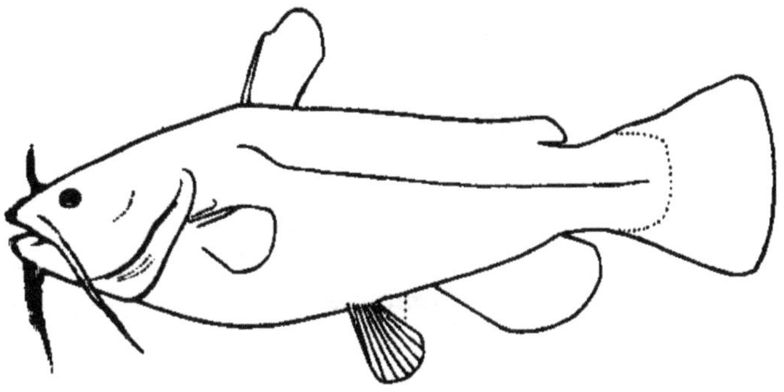

Figure 116: ***Ictalurus melas*** **(Naked Body)**

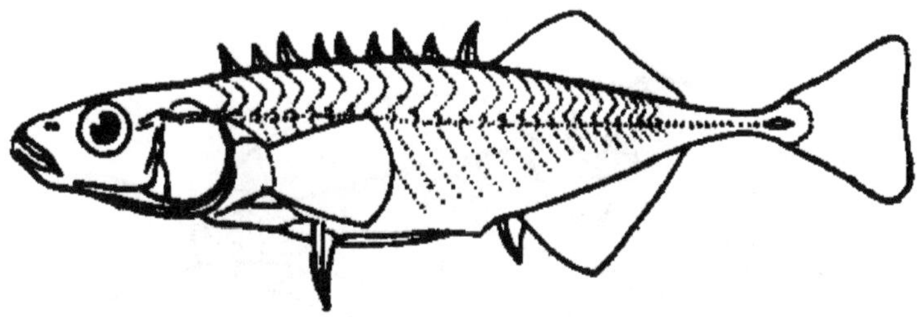

Figure 117: ***Pungitis pungitius sinensis*** **(Naked Body)**

Intermediate

Fishes are nearly naked and have a restricted distribution of scales on the body or scales are reduced to prickles or plates as is the case with *Gasterosteus* species (*Figure 118*), the fish lacks scales but possess bony plates and the extent of bony plating is related to the degree of salinity of water inhabited by fish.

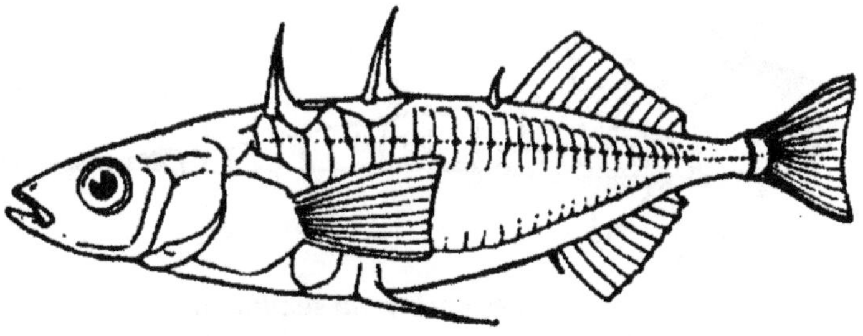

Figure 118: ***Gasterosteus aculeatus***

Similarly, in paddle fishes (*Polyodon* sp.) the skin is naked and scales are restricted to the region of throat, pectoral girdle and upturned base of its tail (*Figure 119*).

In some other cases like *Anguilla anguilla* scales are so deeply inserted that species appear to be naked although actually they are not (*Figure 120*).

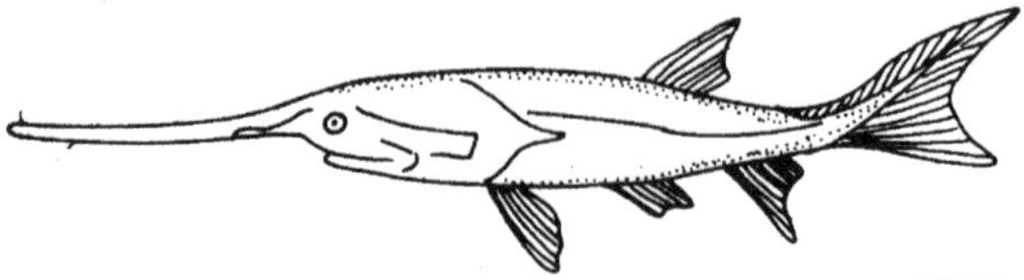

Figure 119: ***Polyodon* sp.**

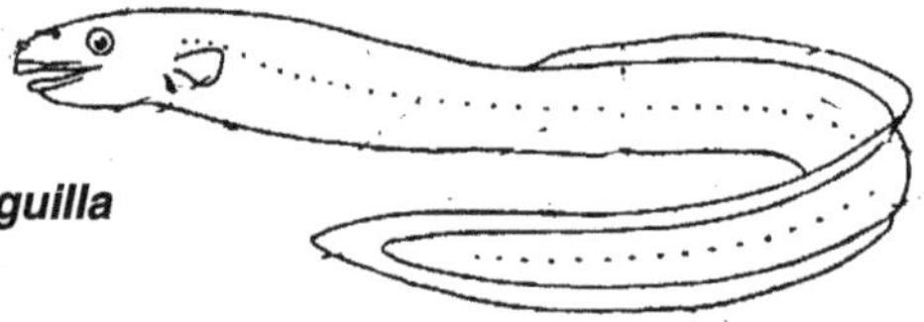

Figure 120: ***Anguilla anguilla***

The sculpins (*fish of northern hemisphere*), inhabiting streams and lakes are naked but variously adorned, the pectoral fins and head are scattered with little prickles that appear to be the vestiges of scales (*Figure 121*).

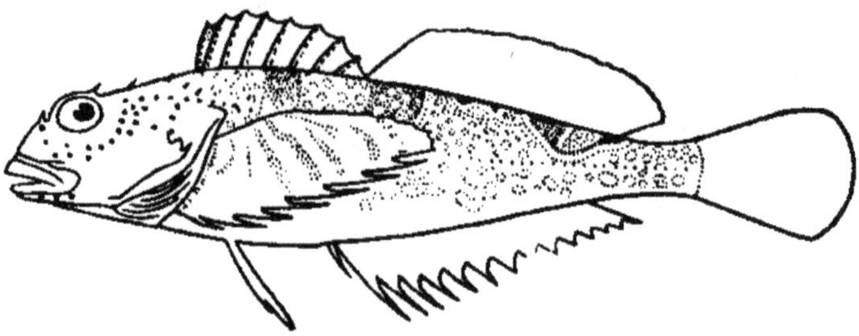

Figure 121: ***Oligocottus maculosus***

In many fish, scales transform into teeth, spiny sting of sting ray (*Dasyatidae, Figure 7a*), bony armouring plates in sea horse (*Figure 34*) and pipe fish (*Figure 35*) and cat fishes of South America (*Figure 122*).

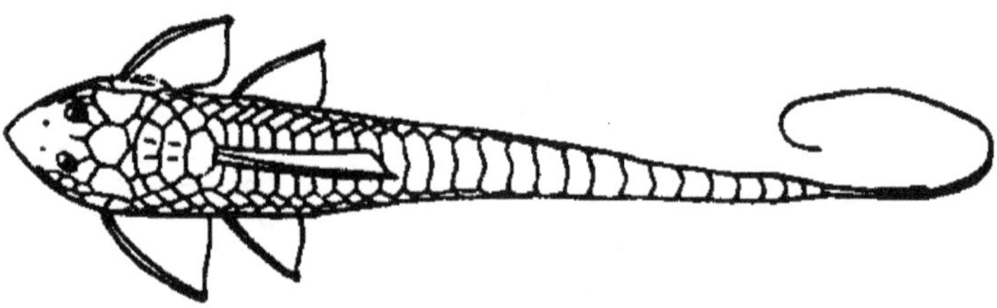

Figure 122: ***Loriacaria seminuda***

Common carp (*Cyprinus carpio specularis; Figure 123*), the mirror variety has a few large scales which are often separated by large areas of clear skin.

Scaly

Scales are thin bony plates covering whole of the body. Their arrangement/ shapes/types/modifications have diagnostic value in fish taxonomy.

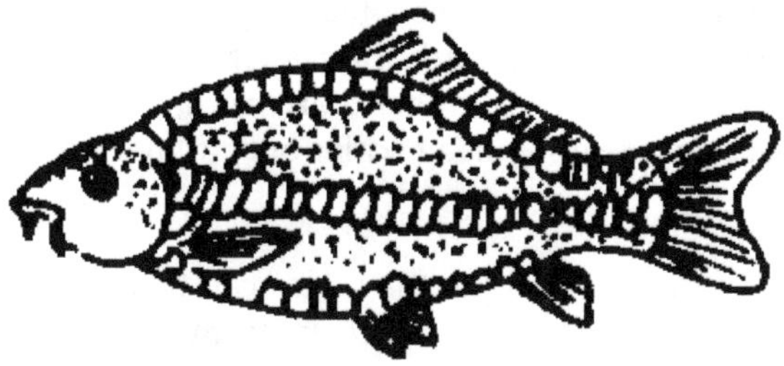

Figure 123: ***Cyprinus carpio specularis***

Shape

On the basis of shape, the scales may be classified as:

Placoid

These are plate-like and each plate carries a cusp *e.g.*, sharks (*Figure 124*).

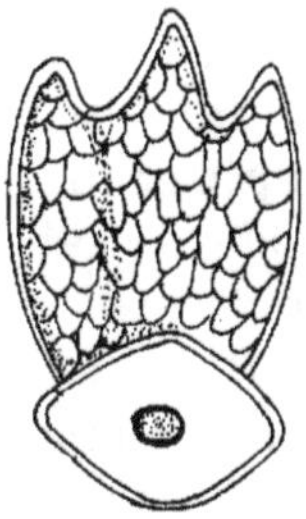

Figure 124: Placoid

Figure 125: Ganoid

Ganoid (Rhomboic)

They are diamond-shaped *e.g.*, reed fishes, gars (*lepidosteidae*) (*Figure 125*). These scales are also present on tail of sturgeons and paddle fishes.

Cycloid

These are disc-like and more or less circular in outline (*Figure 126*). Its exposed caudal margin is smooth *e.g.*, soft-rayed fishes (malecoptergii).

Figure 126: Cycloid

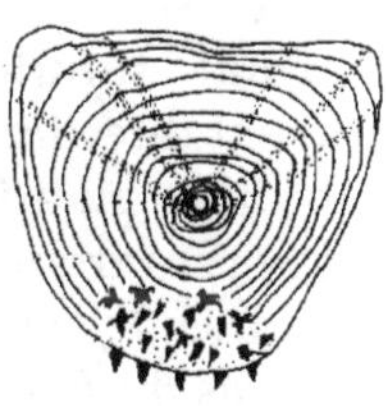

Figure 127: Ctenoid

Ctenoid

In this case the exposed caudal margin is toothed or comb-like *e.g.*, spiny rayed bony fishes (*Figure 127*).

However, some exceptions occur wherein some soft-rayed fishes (characidae, killifishes, cyprinodontidae, poecillidae) have ctenoid scales. Similarly, some spiny rayed fishes have cycloid scales (labidesthes). Whereas many spiny-rayed species (*micropterus*) exhibit both.

Pattern/Arrangement of Scales

Figure 128: Imbrication of Scale

In arrangement, scales are mostly imbricated and overlap like the shingles on the roof (*Figure 128*). Their free margins are directed towards tail in order to minimise the friction with water. This pattern may be totally or partially reversed in some aberrant individuals. In some fishes, like burbot (*Lota lota*) and freshwater eels (*Anguilla anguilla*) the pattern is mosaic and rather than overlapping the scales are minutely separated or meet other neighbouring scales only at margins.

Structure

On the basis of mode of origin, the scales can be categorised as placoid and non-placoid. Placoid scales are those which are formed due to the secretory activity of both epidermis and dermis whereas the non-placoid scales are derived from dermis only.

Placoid Scale

Placoid scales also called as dermal denticles, have a disc-like basal plate embedded in the dermis and a spine projecting out through the epidermis. Structurally it resembles a tooth (*Figure 129*). The spine has an external covering of enamel-like, hard, transparent material called vitrodentine. This is followed by a layer of dentine enclosing a pulp cavity from which several branching dentinal tubules arise in different directions. The centre of the basal plate is perforated by an aperture to provide entrance to blood vessels and nerves from dermis. These scales occur among sharks and their relatives (chondrichthyes). Unlike the scales of bony fishes the placoid scales do not increase in size as the fish grows. In Holocephali, although the skin is naked, in the adult some patches of denticles, essentially similar in structure to those of sharks and rays still remain on the claspers.

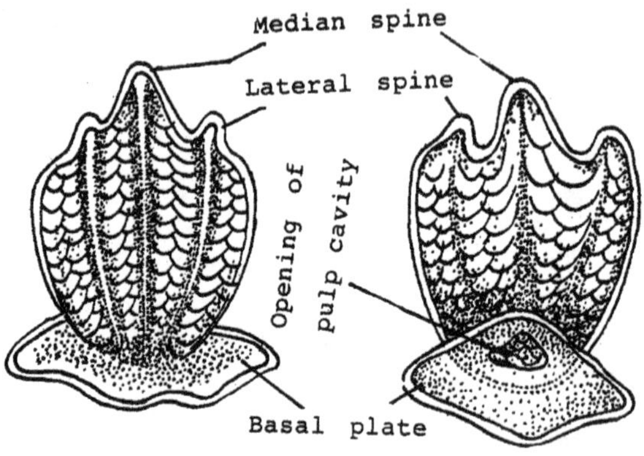

Figure 129: Placoid Scale

Non-Placoid Scale

The non-placoid scales are basically of three types–cosmoid, ganoid and bony ridge.

Cosmoid Scale

In cosmoid scale, the external layer is thin and hard enamel-like and is called the vitrodentine (*Figure 130*). The middle layer is made up of hard, non-cellular, dentine-like material called the cosmine and contains large number of branching tubules and chambers, whereas inner layer is made up of vascularised bony substance, isopedine. These scales, however, grow by the addition of isopedine at the edges from beneath. These scales are found in living (*Latimeria*) and extinct lobefins.

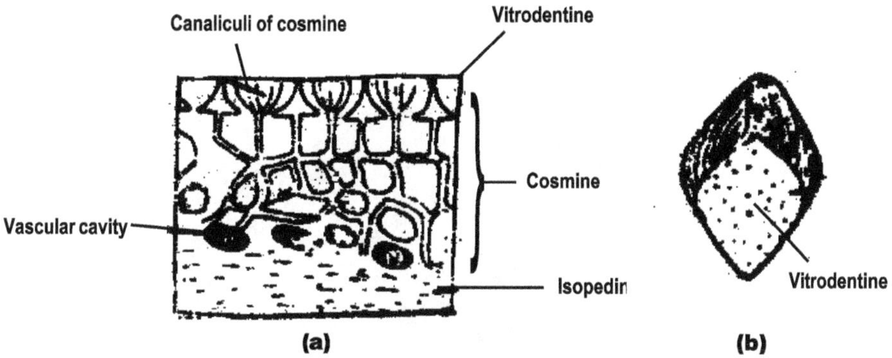

Figure 130: (a) Section of Cosmoid Scale; (b) Cosmoid Scale

Ganoid

In these scales the outer layer is hard with inorganic salt substance called ganoine and differs from vitrodentine of placoids. Beneath the ganoine cap there is a cosmine-like layer and the innermost lamellar bony layer, isopedine (*Figure 131*). These scales grow not only at the edges and from underneath, but also on the surface. These scales are usually rhomboid in shape and articulate by peg and socket joints (*Figure 125*). These scales are best developed in chondrostean and holostean fishes.

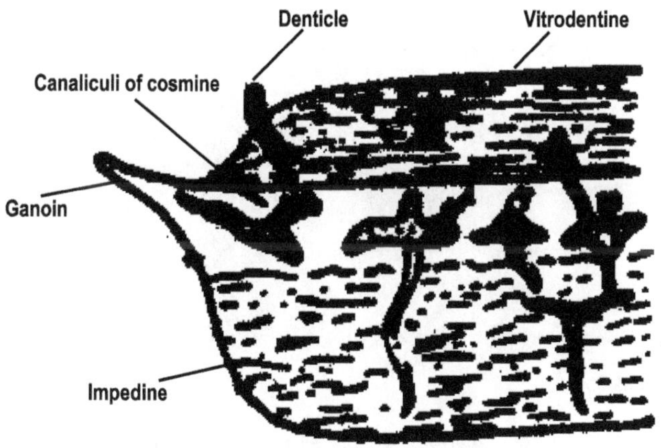

Figure 131: *T.S.* Ganoid Scale

Bony Ridges Scales

These scales are thin and translucent, lack both first and middle layers of other types. Bony ridge scales characterise the many living species of bony fishes which have cycloid or ctenoid scales. The outersurface of such a scale is marked with bony ridges that alternate with valley-like impressions (*Figu*re 132). The inner part of the scale is made up of layers of criss-crossing fibrous connective tissue. The central part of the scale is called focus. The origin and development of both these scales are similar. The cycloid scales are thin and roughly rounded in shape, being thick in the centre and thinning out towards the margins.

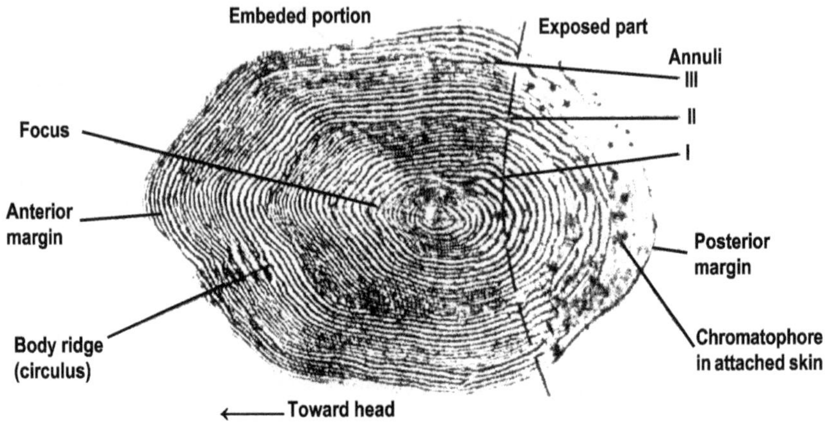

Figure 132: Bony-ridges Scale of the Cycloid Type

The ctenoid scales are also circular (*Figure 133*) and can be distinguished from the cycloid by having a more or less serrated free edges. Moreover, several spines are also present on the surface of the posterior area of the scale.

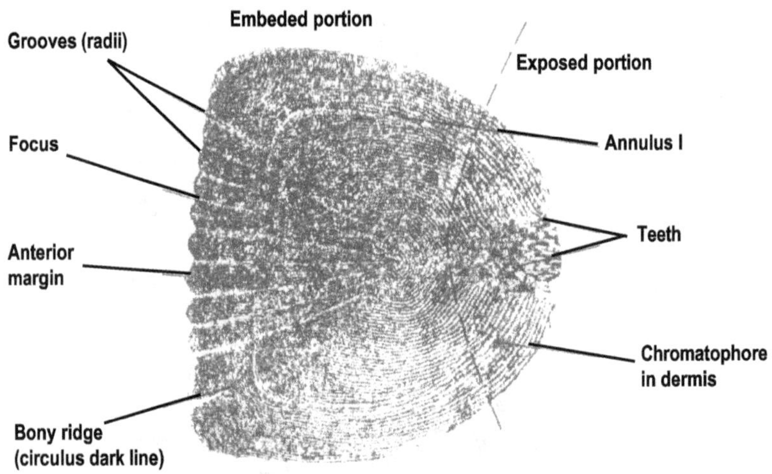

Figure 133: Bony-ridge Scale of the Ctenoid Type

Derivatives of Scales

Jaw Teeth

In sharks and their relatives jaw teeth are the modifications of placoid scales (*Figure 134*).

Spines

Scales are modified as spine on dorsal fin of spiny dog fish (*Figure 5*) and chimaeroids (*Chimaera* sp., *Hydrolagus* sp.) (*Figure 9*).

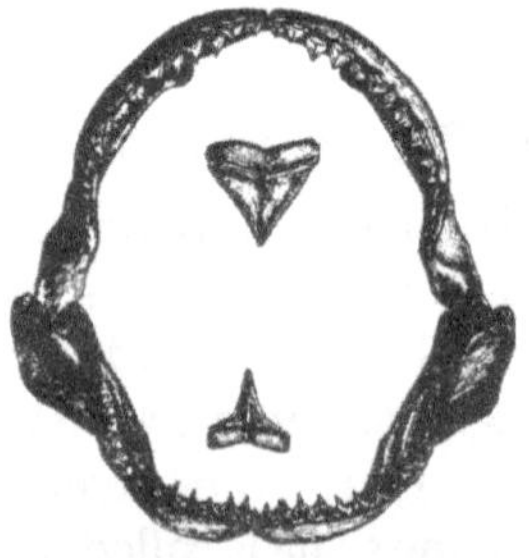

Figure 134: Dentition of the Blue Shark (*Carcharias glaucus*) (After Gunther, 1994)

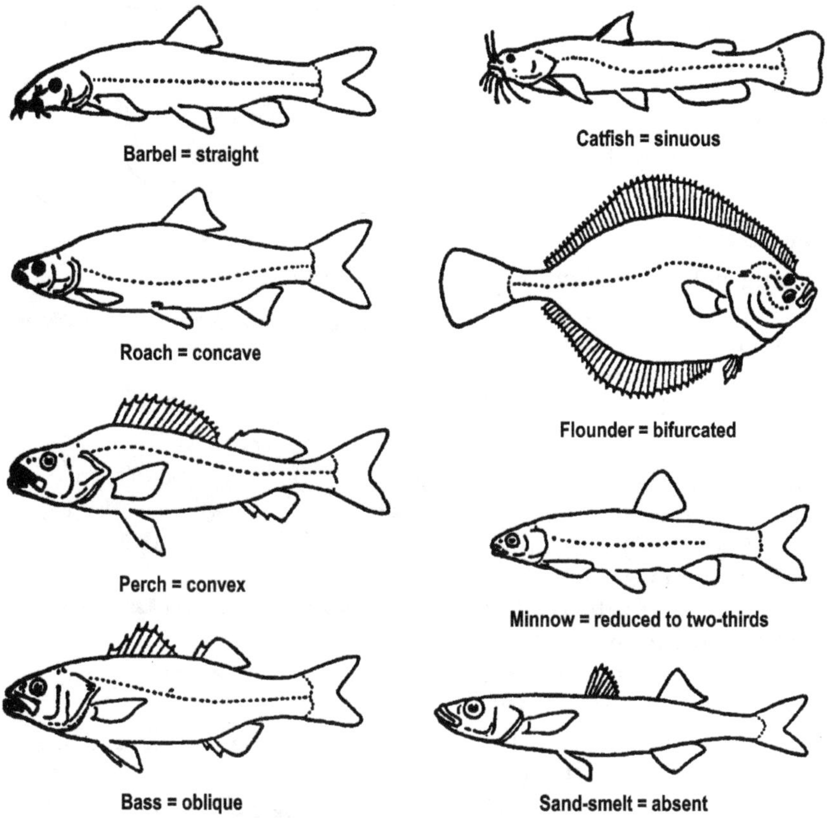

Figure 135: Profile of the Lateral Line (After Arrignon, 1999)

Stinger

On the tail of sting ray, scale is modified as stinger (*Figure 7a*).

Saw Teeth

Saw teeth of pristis are modified scales (*Figure 7b*). Lancets of sturgeons originate from bony ridge scales (*Figure 12*).

Lateral Line

Special perforated and tubular modifications of scales constitute a channel for lateral line sensory canal in many fishes. This lateral line is composed of large number of sense organs called neuromasts, each of which consists of a group of sensory cells with hair-like projection. Lateral line system provides 'distant touch sense' of objects. The lateral line shows various shapes in different species. It may be straight, sinuous, concave, bifurcated, convex, reduced to two-third, oblique or absent (*Figure 135*).

Supraorbital

Running above eye and reaches upto snout (*Figure 136*).

Infraorbital

Running below eye to the snout (*Figure 136*).

Hyomandibular

Passing behind the eye and going to the lower jaw. The hyomandibular canal is situated on outside the hyoid region and hence is prolonged forward and downward (*Figure 136*).

Most of the fishes have lateral line but in some it is absent (muglidae) (*Figure 137*). This lateral line may be continuous (*Figure 138*), extremely short or interrupted (*Figure 139*) or present on dorsal side or ventral side, or discontinuous with two levels (*Figure 140*). Generally, lateral line stops at the base of caudal fin but in *Lates calcarifer* it extends beyond into caudal fin (*Figure 141*).

Axillary Lobe

Enlarged scale at the base of ventral fin (*Figure 99*).

Alae

Scale-like appendage at the base of caudal fin (*Figure 98*).

Bony Plates

Rhomboid plates arranged in rows on the body (*Figure 142*).

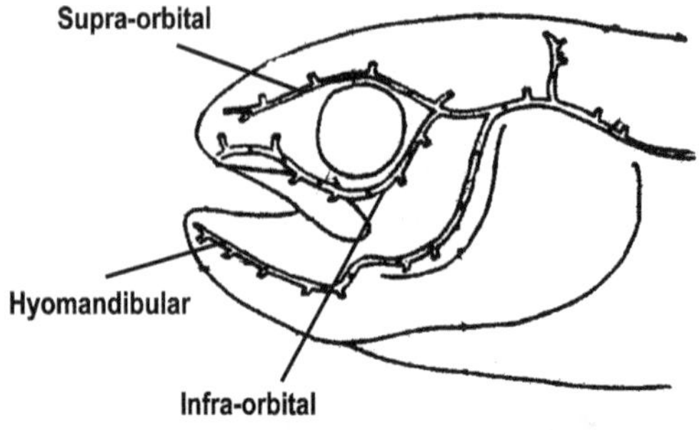

Figure 136: Different Locations of Lateral Line

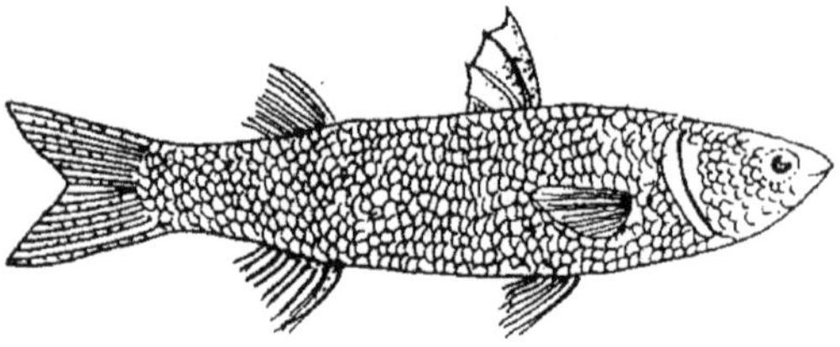

Figure 137: Lateral Line: Absent

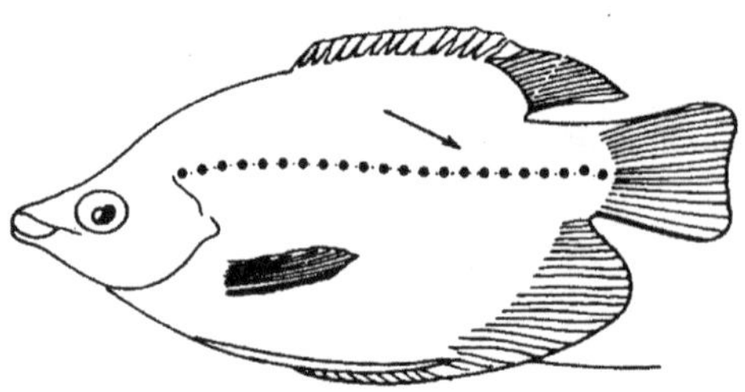

Figure 138: Lateral Line: Continuous

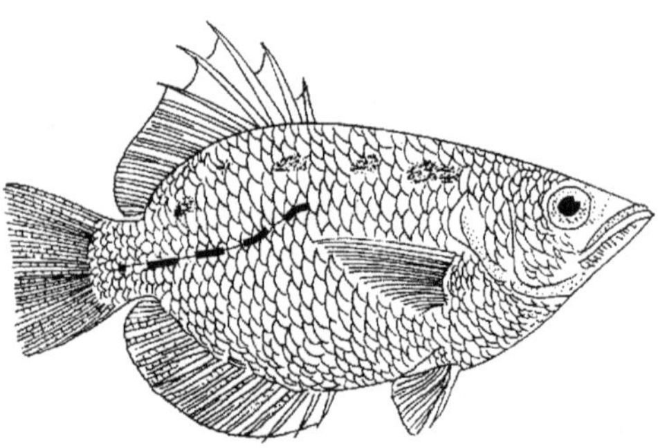

Figure 139: Lateral Line: Short

Use of Scales in Classification

Fish scales are useful for ichthyologists for classification. Major groups of living fishes (lampreys and hag fishes) are scaleless, sharks and their relatives (chondrichthyes) are characterised by placoid type of scales, lung and lobe fin fishes (sarcoptergii) have cosmoid scales, and primitive bony fishes oftenly have bony ridge scales.

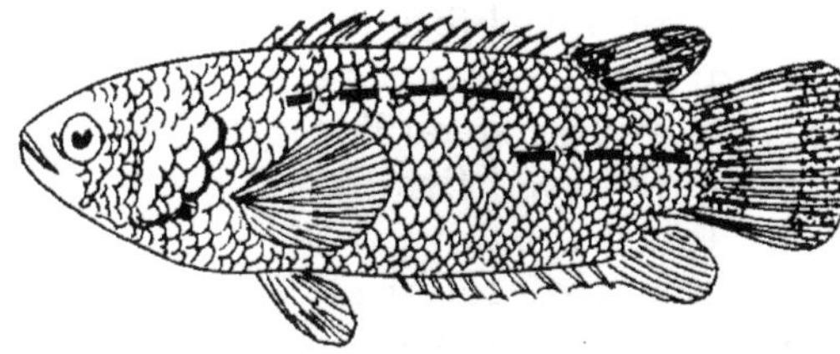

Figure 140: Lateral Line: on Dorsal Side

Besides their use in fish classification, scales are also useful in separating orders and families. Scales have utility too in classifying the remains of fishes in the waste heaps of prehistoric man and in fossil deposits.

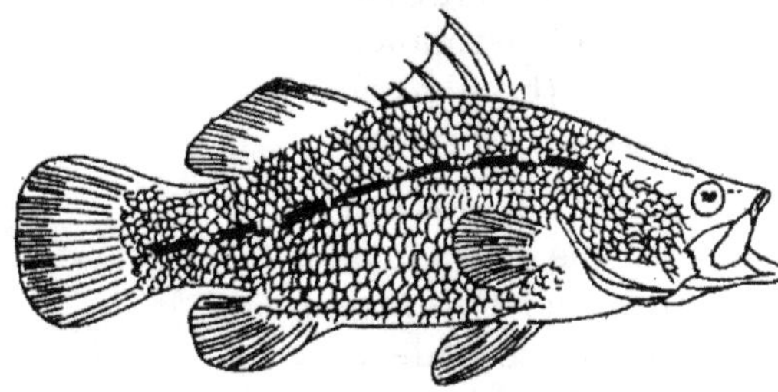

Figure 141: Lateral Line: Extending into Caudal Fin

The diagnostic value of scale morphology for the separation of species varies among families or genera. In taxonomic work, highly useful is the scale count, their number in the lateral line, and rows along or around the body.

In addition to identification and classification, scales also interpret the life history, for example in determining the age of fish from annuli on the scales. From the spacing of annual rings and a knowledge of the length of individual at capture, fish length at each previous year mark may be calculated after proportionality of scale growth to body growth has been determined. Sometimes scales also disclose how many times a fish has spawned. In some fishes scales have spawning marks (Atlantic salmon) in the form of rings which are the result of cessation of feeding and exhaustion during the spawning period.

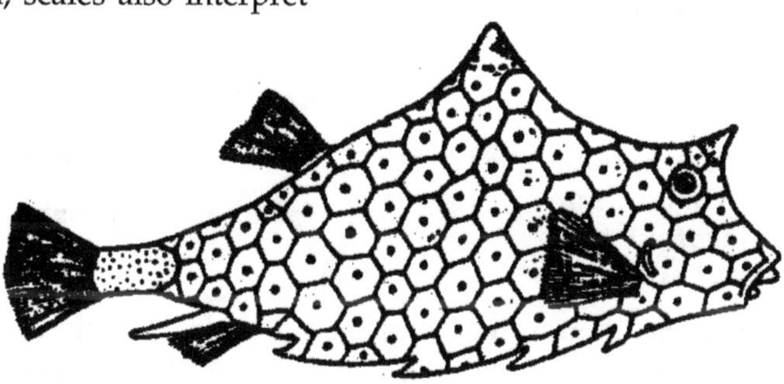

Figure 142: ***Ostracion gibbosus***

Scale Reading and Age Determination

There is a definite way of expressing the scale count. If the scale formula is given $100^6/8$, it means that there are 100 scales in the lateral line, six scales above it and eight scales below it (at the level of base of dorsal fin/at a point fish have maximum depth).

For the age determination in soft-rayed fins, the scales are taken from the region above the lateral line but in spiny-rayed fins the scales located below the lateral line are preferred. Large and symmetrical scales are taken for age determination. The scales are cleaned in water and permanently mounted in the glycerine-gelatin mixture for study.

The true annulus can be marked on the scale as a thick and prominent lining in between the considerable number of ridges and grooves. The total number of annuli on the scale indicate the age in same number of years (*Figure 143*).

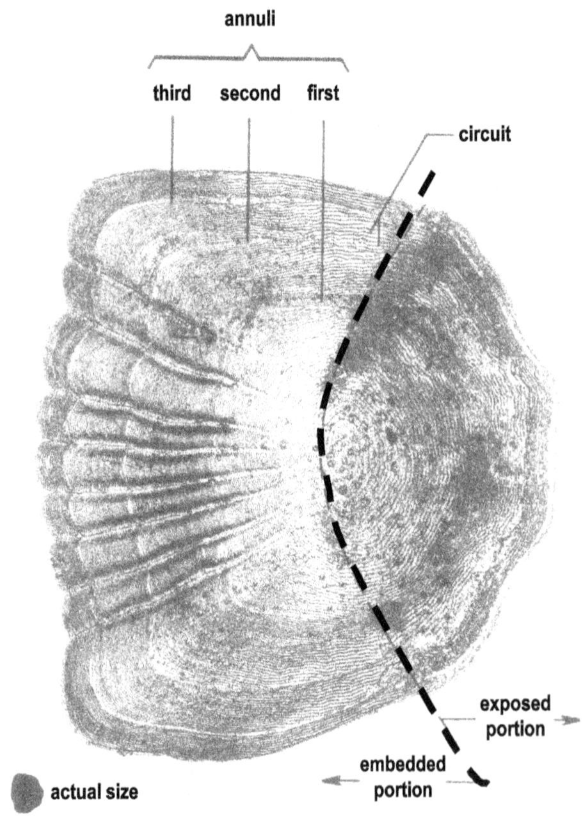

Figure 143: True Annulus

Coloration

Certain generalisations can be drawn from a large variety of coloration that fishes exhibit as a basic pattern.

Freely swimming, open water fishes are of simple coloration grading from a whitish belly, through silvery lower side, to upper sides and back that are irridescent blue or green.

Bottom dwellers and weed-bed occupants are strongly marked above and pale beneath. Three types predominate in colour of oceanic fishes: silver in the upper zones (*i.e.*, fish occuping pelagic zone); red in the middle range (*i.e.*, fish occupying bathyal zone) and violet or black in the greater depths (*i.e.*, fish occupying column).

Mostly the freshwater fishes have red markings while the marine have yellow.

In clear and turbid water distinction also occurs. Fish inhabiting clear water tend to have blue or green and those living in turbid water tend to have yellow, orange or red colours or even blue, green, gold and red.

Fish present coloration in myriad shades, so more sophisticated scale is needed to measure it. Nowadays, fish coloration is measured in terms of Munsell colour system which is based on not the characteristics of fish visual (colour) system but on those of human visual colour system.

The coloration in fishes is due to chromatophores and iridocytes. The principle functions of coloration are–concealment from enemy (camouflage), means to catch the prey by disguise or means to advertise the presence of the fish (for sex recognition) (*Figure 144*).

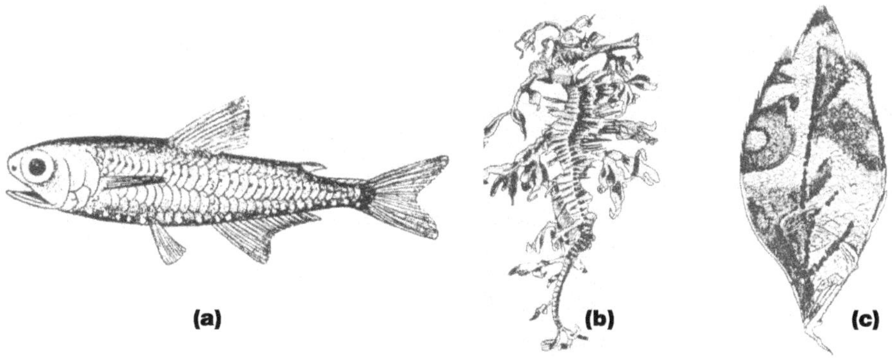

Figure 144: (a) Lanternfish (*Myctophum* sp.); (b) Sea Dragon (*Phyllopteryx* sp.); (c) Leaf-mimicking Fish (*Monocirrhus* sp.) (After Lagler *et al.*, 2003)

Colour Pattern

Some of the most common colour patterns are:

Cryptic Colouration

In order to be invisible many fishes are coloured to match their back-grounds or are coloured to match their background (*Figure 145*). This cryptic colouration is particularly important for sluggish bottom fishes. Where they match very closely to the bottoms on which they rest. Some species of tide pool fishes may be green when living among green algae fronds but change to red when they live among red algae.

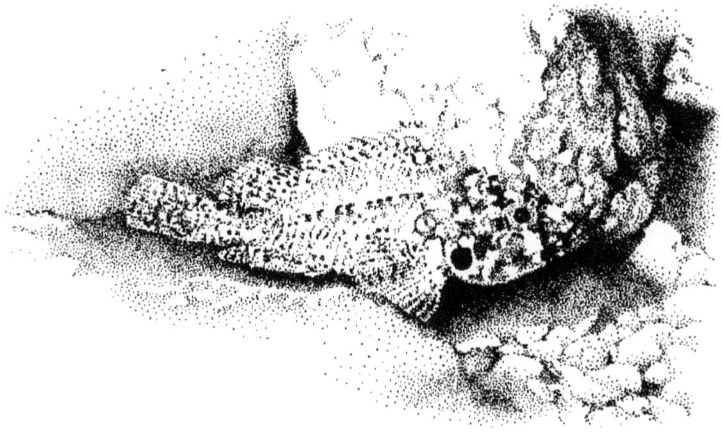

Figure 145: Cryptic Coloration (Rainbow scorpionfish) (After Moyle, 1993)

The flounders are the real masters of change which change their colour pattern to match the bottom they are resting on.

Silvery

The most characteristic of well lighted waters are bright silvery fishes (*Figure 146*). The individuals become very hard to pick out due to the light flashing off the scales. This silvery colour also helps the fish to blend in the light reflected from the surface of water. Trouts and salmons possess complex colour pattern in the streams and change to silvery in the lakes or oceans.

Figure 146: Silvery Appearance of School of Pacific Sardine (After Moyle, 1993)

Countershading

Fishes living in the water column have dark backs and white bellies (*Figure 147*). This dark back helps them to blend with the dark bottoms when viewed from above, whereas the white belly helps them to blend with the sky when viewed from below.

Figure 147: Countershading (Batray)

Disruptive Colouration

The colours and patterns breakup the outline of fish and makes them harder to be seen (*Figure 148*). The most common pattern seen is vertical bars running down each side and such pattern is associated with fishes living near the beds of aquatic plants. Such vertical bars on the fish can blend in with the vertical pattern of the plant stems and thus confuses the predator.

Red Colouration

Most red fishes in the ocean live below the depths of red light penetrations or are active only at night and are difficult to be seen. In shallow water, many fishes such as stickle backs and trout, use red as a breeding colour. In most of the fishes red is confined only to the males and is developed only during breeding season.

This colour makes them less vulnerable to the predators and also makes them more attractive to females.

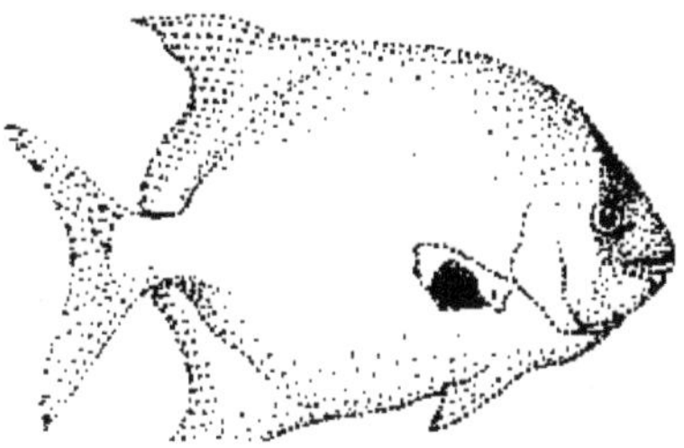

Poster Colours

Many coral reef fishes have complex colour patterns (*Figure 149*). The main functions of which are: (*i*) to advertise the territories of the owners; (*ii*) signal to predator that the possessor is too spiny or poisonous to be worth eating; (*iii*) important signs of sex, status or maturity; (*iv*) camouflage.

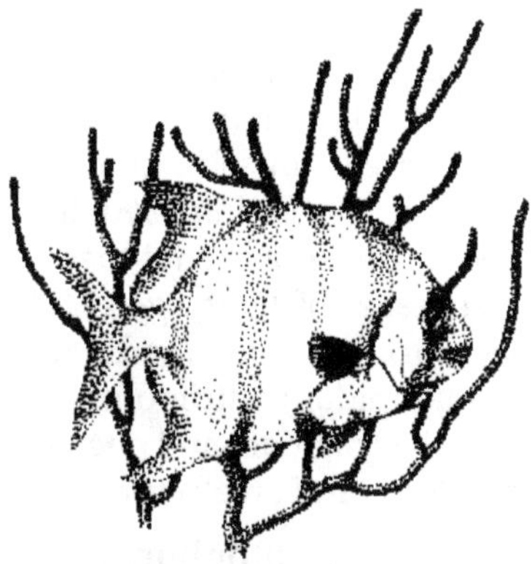

Figure 148: Disruptive Coloration (Pacific Spadefish) (After Moyle, 1993)

Eye Ornamentation

Eyes are the focus of attacks by predators and are important in communication with other members of the species. This results in two contrary trends in eye ornamentation; one is to disguise the eyes and other is to emphasise them. The way of disguising eye is that a black line runs through the pupil and is continuous with either horizontal or vertical stripes on the body, the contrast between iris and pupil is also minimised by having former dark in colour as well as the area surrounding the eye, having numerous marks surrounding the eye that are similar in size to the pupil. The ways to emphasize the fish eyes are that they are brightly tinged with blue, green or yellow colours. Supplementary patterns like eye rings are also common in reef fishes.

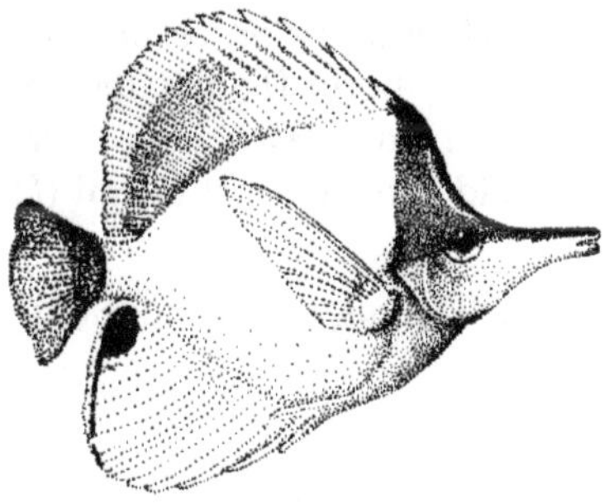

Figure 149: Poster Colours (Butterfly Fish) (After Moyle, 1993)

Eye Spots

Some juvenile fishes have black spot near the base of the tail and is about the size of the eye. It may be emphasized with a light-coloured ring while the real eye is disguised. The principal function of this spot is to confuse the predator, by having them aim for the tail rather than head, giving victim a greater chance to get away. In one of the species of South African cichlids, the eye spots of juveniles keeps away the parents from eating them.

Lateral Bands

In schooling fishes, single dark band runs along the side of a fish. This band keeps the members properly oriented to one another. These bands also confuse the predator because bands in a school may blend together and make it difficult for the predator to pick out one individual.

Light Organs

Number of fishes, mostly seaforms living at a medium depth of 500-2500 m or deep water forms possess light producing organs/luminiscent organs called photophores. These are specialised glands of epidermis. Its biological significance is unknown. Systematists, however, have found use of organs as taxonomic characters, *e.g.*, in the lantern fishes and relatives (myctophiformes).

The luminiscence in fishes is of two types: (*a*) that which results from the presence of luminous bacteria living on the fish in a symbiotic relations; and (*b*) that which arise from self-luminous cells on the fish, the photophores. The luminous bacteria on the fish inhabit definite pockets in the skin, often backed by the mirroring tissue and externally equipped with a shutter arrangements of some kind to regulate the emission of light *e.g.*, fish families with bacterial light origin are grenadiers (macrouridae), cods (gadidae), pinecone fishes (monocentridae), anomalopids (anomalopidae), slipmouths (leiognathidae), basses (serranidae), swallowers (saccopharyngidae) and deep sea angler fishes and some relatives of lophiiformes, etmopterus and an electric ray.

The photophores show considerable variation in their number and mode of distribution on the body. They may occur along the lateral line or ventral sides of the body or head region. They occur on certain limited parts of the body (*Figure 150*) or are arranged in one or two rows extending on the sides from head to tail (*Figure 151*).

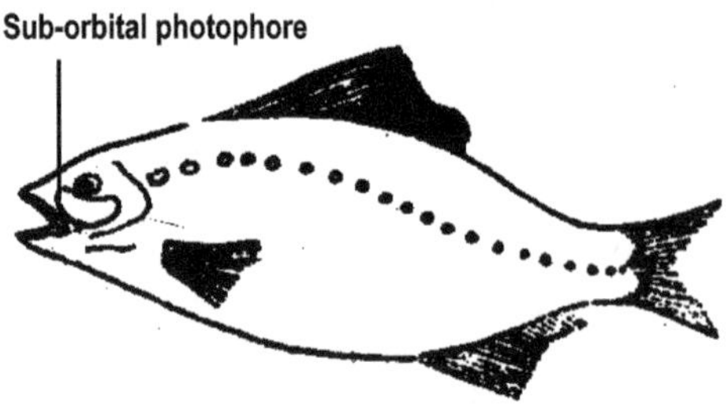

Figure 150: *Photoblepharon* sp. (Sub-orbital Photophore)

The luminous barbel under the chain has been reported in *Macrostomias* (*Figure 152*).

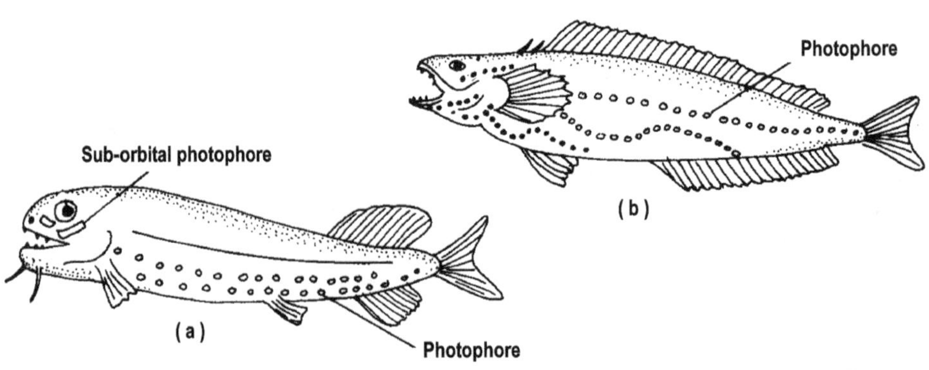

Figure 151: (a) *Pachystomias microdon*; (b) *Porichthys* sp.

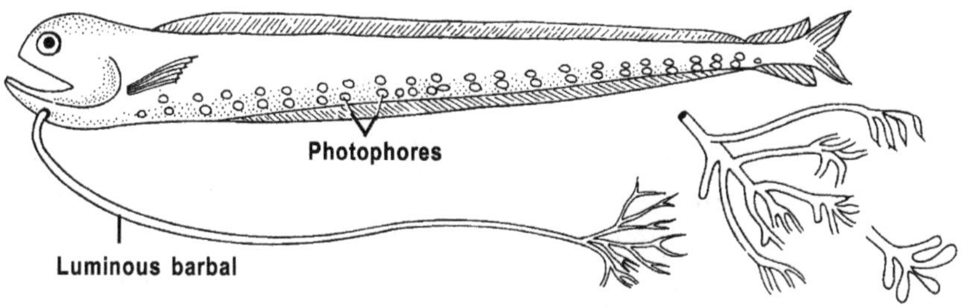

Figure 152: *Macrostomias* sp.
(After Khanna and Singh, 2003)

The anatomy of light producing organs can be described in detail under several headings as:

Simple Organs

The simple photophores consist of a series of radially arranged glandular tubules which receive branches from adjoining cranial or spinal nerves. Their size is very small, not more than 0.1 to 0.3 mm in width. Simple organs are with or without an interrupted mantle of pigment. Grouping of cells in photophores are also called as 'lenses' or 'lenticular cells'. The distal area of the photocytes contains acidophilic granular material. A layer of melanophores invests the photophores.

Compound Organs

In these organs in addition to the photogenic tissue, a reflecting layer and a lens-like structure is also developed. The sub-orbital organ of Pachystomias is also complex in structure (*Figure 153*). It is cup-like and wall is composed of concentric layers. Externally, there is a layer of black pigment in the cup and internally it has numerous tubules. A thick layer of light reflecting spicules is present in the cup whose axial part is full of a number of radially arranged gland tubules. The mouth of cup is occupied by a lens-like structure and skin forms a covering like an iris diaphgram over it. This organ is supplied by a branch of fifth cranial nerve. Both simple and complex type of photophores may be present in the same species.

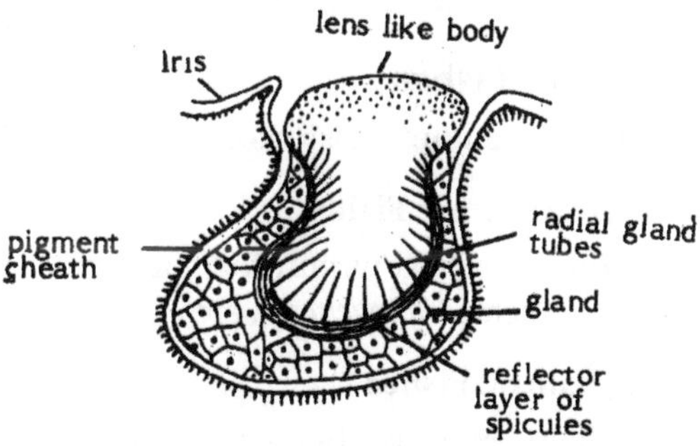

Figure 153: Section of sub-orbital photophore of *Pachystomias*

The luminescent glands consist of luciferin and luciferase (enzyme) which when combined produce light. Luciferin is an indole derivative consisting of tryptamine, arginine and isoleucine. When mixed with the enzyme luciferase, luciferin is oxidised to oxyluciferin, emitting light. In fishes, luminiscence is generally blue or blue-green.

The light producing organs are directly under the control of nervous system and endocrine glands.

Uses

The most important use of this light is to provide light in places of darkness as in deep sea where ordinary sunlight does not penetrate. They are also used as a defensive organ to confuse enemy for a moment, enabling the fish to escape. Some fishes make use of their luminiscent organ to attract the prey near the mouth. Luminiscence in Porichthyes serves as a warning signal for the predatory animal. The pattern of photophores on the body is also useful in recognising individuals of same species.

Poison Glands

Poisonous glands are present in many fishes as skin derivatives. These are epidermal glandular cells which may or may not be isolated from rest of epidermal cells. They secrete venom which is somewhat like that of snakes and is a frequent cause of death when it is injected through the spine.

About 1000 species of fishes are thought to be venomous or poisonous. The term "poisonous fishes" is applied to those fishes whose tissues either in part or entirely are toxic.

The term "venomous fishes" is limited to those fishes which are capable of producing a poison in highly developed secretory organ or group of cells and which can deliver this toxin during a biting or stinging act.

Poisonous Fishes

There are four divisions of poisonous fishes.

Ichthyosarcotoxic Fishes

Where toxin is within their musculature, viscera or skin *e.g.*, cyclostomates, elasmobranchs, *Chimaera* sp. etc.

Ichthyootoxic Fishes

Where toxin is confined only to female gonads. In most of the freshwater species the roe is poisonous *e.g.*, *Schizothorax* sp.

Ichthyohaemotoxic Fishes

Such fishes have toxin in their blood *e.g.*, some freshwater eels and some marine fishes.

Ichthyocrinotoxic Fishes

Such fishes produce toxin by glandular secretion, but lack a true venom apparatus, these produce toxic substance in their skin and subsequently release into their environment.

Venomous Fishes

These are frequently found in the Pacific areas and are inshore fishes. Some are found in shallow waters and some are found in deep waters. Venom from sting of following fishes may be fatal as well as having painful symptoms to man. Few examples of venomous fishes are:

Sharks

Fine spine in spiny dog fish (*Squalus acanthias*) and groove-horn shark (*Heterodontus francisci*).

Rays

Caudal stinger in sting ray (*Dasyatis* sp., *Figure 7a*), spotted eagle ray and bat sting ray (*Myliobatis californicus*).

Chimaeras

Dorsal fin spine in *Chimaera* sp. (*Figure 9*) and *Hydrolagus* sp.

Scorpion and Rock Fishes (*Scorpaenidae*)

Dorsal, anal and pelvic fin spines in scorpion fishes (*Scorpaene* sp.), zebra fishes (*Pterois* sp.) and stone fishes (*Synanceja* sp.).

Weaver Fishes

Opercular stinger and spines of dorsal fin–*Trachinus* sp.

Toad Fishes

Opercular stinger and dorsal fin spine–in *Barchatus* sp., *Thalassophryne* sp.

Cat Fishes

Species of dorsal and pectoral fin in stonecats, madtoms (*Noturus* sp.), *Ictalurus* sp., *Heteropneustes* sp., *Clarias* sp.

Sturgeons

Stinger on each side of caudal peducle *e.g.*, *Acanthurus* sp.

Dragonets

Fin spines with venomous glands–*Callionymus* sp.

Rabbit Fishes

Dorsal, anal and pelvic fish spines–*Siganus* sp.

Stargazers

Shoulder stinger.

Derivatives of Skin

The skin may get extended into barbels and flaps.

Barbels

Some fishes have "whiskers" around their mouth called as barbels. These are accessory feeding structures that carry sense organs. Their number and size varies considerably (*Figure 154*). In the fishes that feed mainly on bottom invertebrates the barbels may become longer. In some sheath-fishes there may be as many as 16 barbels

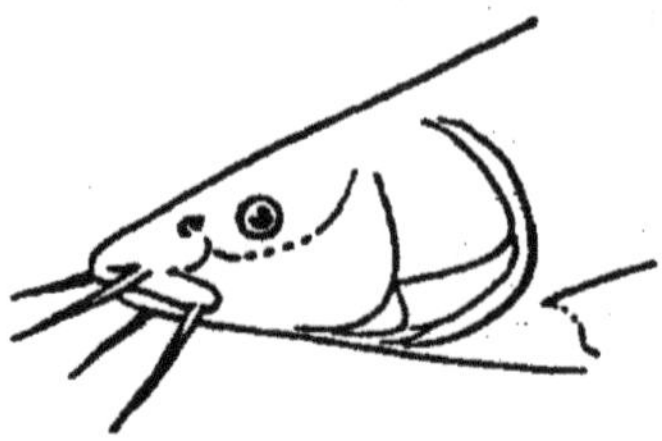

The barbel (*Barbus barbus*)

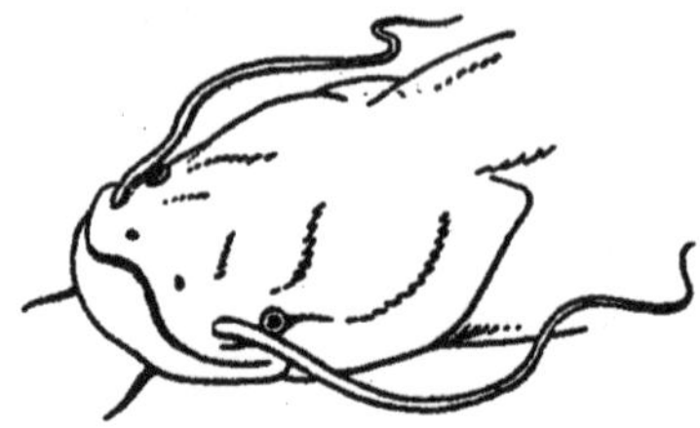

The sheat-fish (*Siluris glanis*)

The cod (*Gadus morhua*)

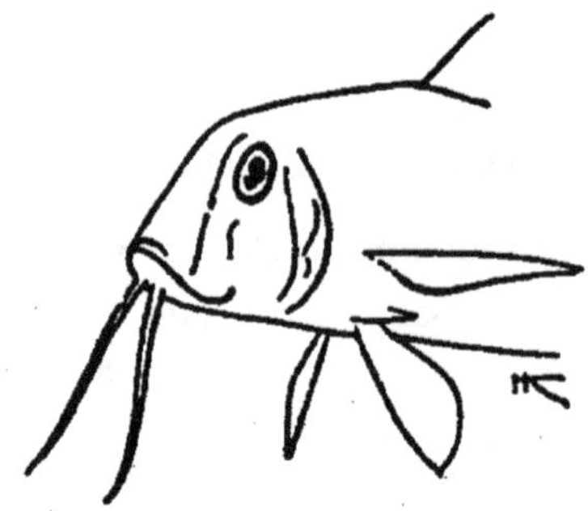

The red mullet (*Mullus barbatus*)

Figure 154: Various Types of Barbels of Fishes

and some of these may be even longer than the body (*e.g.*, Indo-Malayam sheath fish, *Siluorides eugeneiatus*).

In mormyrids, there is an outgrowth at the end of the snout which is of considerable length and rich in sensory nerve endings. By means of this snout it burrows in the silt and feels the buried invertebrates on which it feeds (*Figure 155*). The barbels may be soft and muscular (*Figure 156*), stiff and osseous (*Figure 157*) or with simple hollow tubes (*Figure 158*) or annulated (*Figure 159*). When branched they look like weeds attached to the chin.

(a)

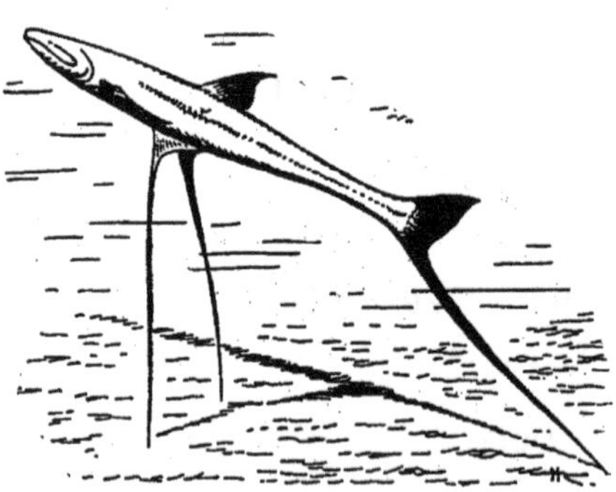

(b)

Figure 155: (a) The Gourami (*Osphoronemus* sp.); (b) *Benthosaurus* sp. (After Nikolsky, 1999)

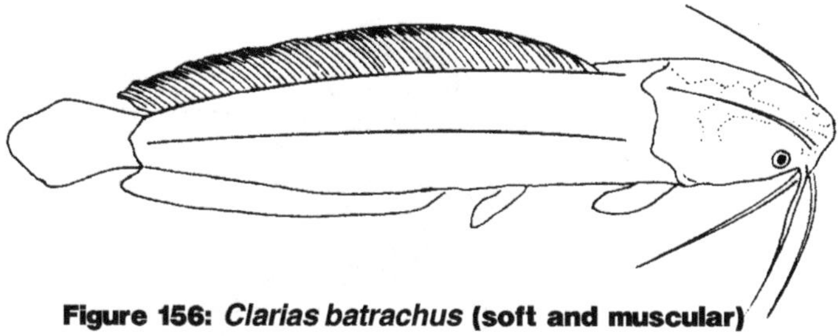

Figure 156: ***Clarias batrachus*** **(soft and muscular)**

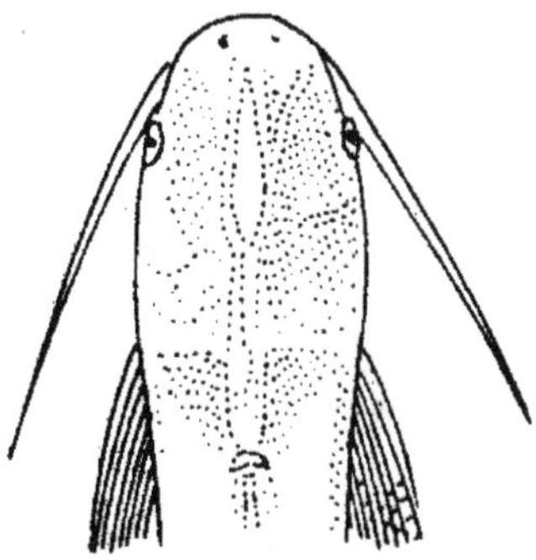

Figure 157: ***Osteogeniosus militaris*** **(Stiff and Osseous)**

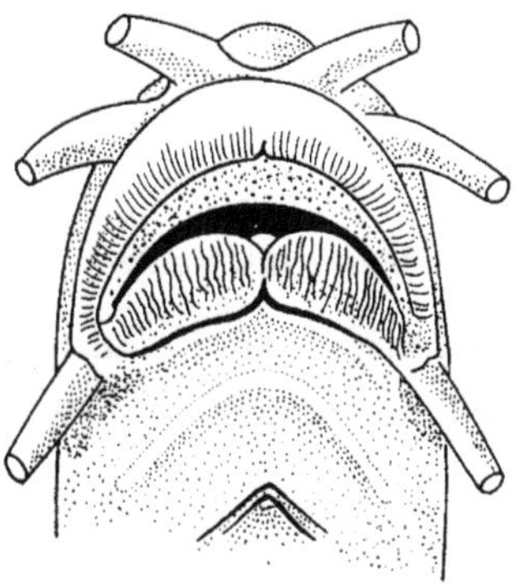

Figure 158: ***Noemacheilus labeosus*** **(Simple Hollow Short Tubes)**

Depending upon their position, barbels may be:

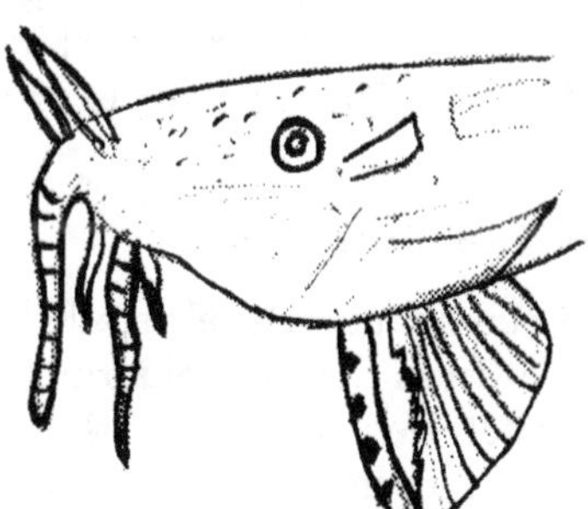

Figure 159: Barbel–Annulated

Hyoid Barbels

When attached in the interbranchial space in the region of hyoid arch (*Figure 160*).

Maxillary Barbel

When attached to the maxillary part (*Figure 161*).

Mandibular Barbel

When arising from the mandibular region (*Figure 161*).

Figure 160: Hyoid barbels (*Polymixia japonica*) (After Lindberg, 1976)

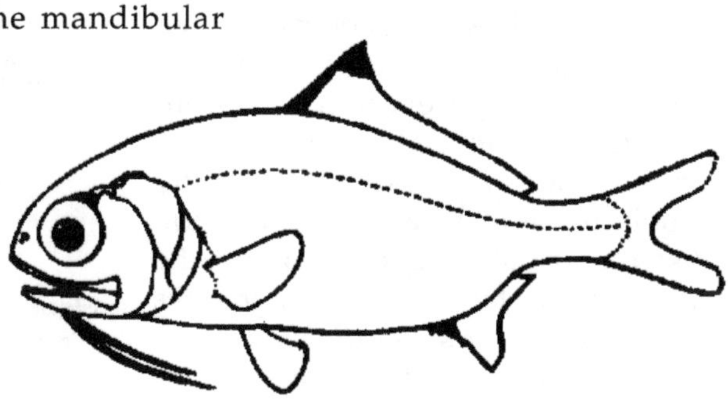

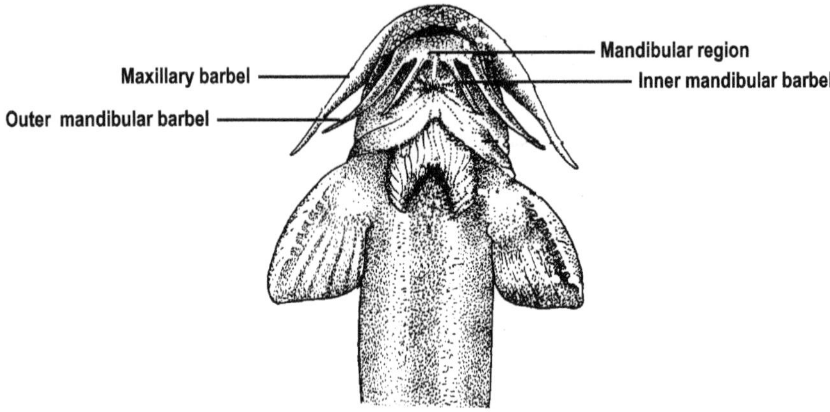

Figure 161: Ventral View of Anterior Region of *Glyptothorax*

Inner and Outer Mandibular Barbel

When they originate from the inner and outer sides of the mandibular region (*Figure 161*).

Rostral Barbels

When arise from rostrum (*Figure 162*).

Figure 162: Rostral Barbels

Flaps

In sargassum fish and algae resembling seadragon (*Phyllopteryx*), the skin gets extended into flaps and its function is protective (*Figure 144b*).

Breeding Tubercles

Breeding tubercles also called as pearl organs are local keratinizations distinctly different from normal epidermis. These are epidermal structures which facilitate contact between individual fishes during spawning. These are present in 17 families of bony fishes in 5 orders: salmoniformes, gonorhynchiformes, cypriniformes, scorpaeniformes and perciformes.

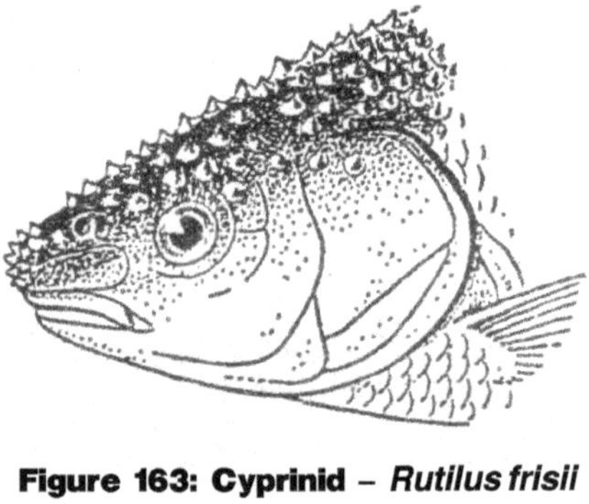

Figure 163: Cyprinid – *Rutilus frisii kutum* with nuptial array. (After Nikolsky, 1999).

There are three types of roughening structures (nuptial tubercles): tubercles consisting of aggregations of non-keratinized epidermal cells; tubercles containing substantial number of fully keratinized cells that are organised to form a discrete, usually conical cap and the contact organs which are bony outgrowths from the fin rays or scales (*Figure 163*). In most of the cases, these are present only on the male or are much developed in males as compared to females. They begin to develop before the spawning season, reach their maximum contact just before or during the spawning season and then regress.

Breeding tubercles develop only on freshwater or inshore marine spawning species.

Adhesive Organs

Various sort of adhesive organs are present in fish for attachment, adhering or mechanical grasping to the substratum. These may be suction cups, sticky pads or hooks. Many fishes have adhesive organs formed from modified lips or thoracic fins. Adhesive structures are integumentary modifications and vary in location, form and size.

The skin of ventral side of outer rays of paired fins is greatly thickened to form a cushion like structure that helps in attachment to rocks and stones. In few (cyprinids) fishes, skin surrounding the mouth becomes thickened and tuberculated for attachment.

In a few the skin is thrown into grooves and ridges *e.g., Glyptosternum* sp., *Glyptothorax* sp., *Pseudoecheneis* sp. etc.

In a few cases the lips becomes thick and horny, and function as sucker, *e.g., Nemacheilus* sp. (*Figure 164*).

In a few, jaws are covered by a hard callous plate for scrapping the food. Posteriorly the plate gets thickened and tuberculated *e.g., Schizothorax* sp. (*Figure 165*).

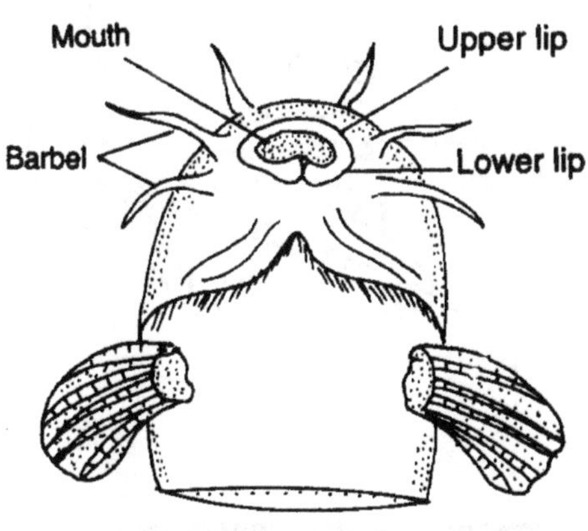

Figure 164: Ventral Side of Head of *Nemacheilus* sp.

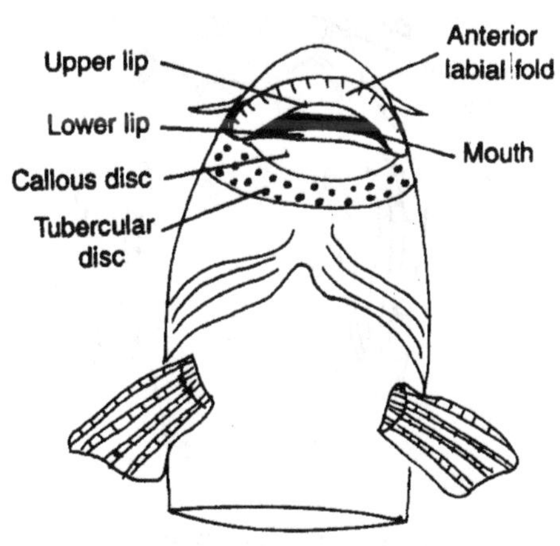

Figure 165: Ventral Side of Head of *Schizothorax* sp.

In a few the adhesive organ in the form of disc is present on posterior-ventral side of mouth and consists of central callous part and free tuberculated lateral and posterior borders *e.g., Garra* sp. (*Figure 166*).

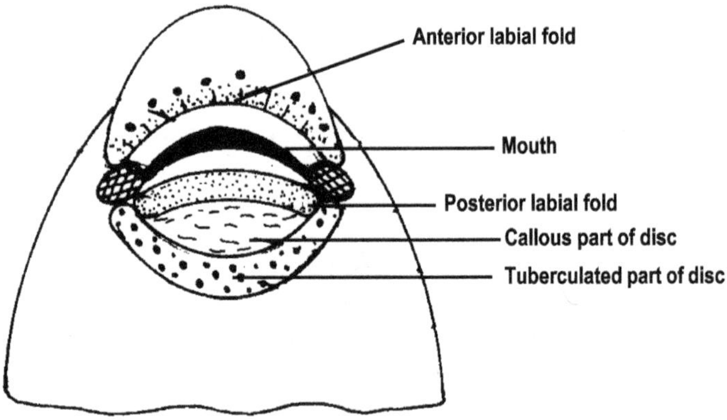

Figure 166: Adhesive Apparatus of *Garra gotyla*
(After Khanna and Singh, 2003)

Sometimes striations are present on the chest and belly that help in adhesion *e.g., Laguvia* sp. and *Erethistes elongata.*

In *Glyptosternum* sp., the skin on the under surface of the pectoral spine is striated and each ridge is supported by short pointed cartilaginous ray given out from the outer margin of first pectoral and pelvic rays *e.g., Glyptosternum* sp. (*Figure 167*).

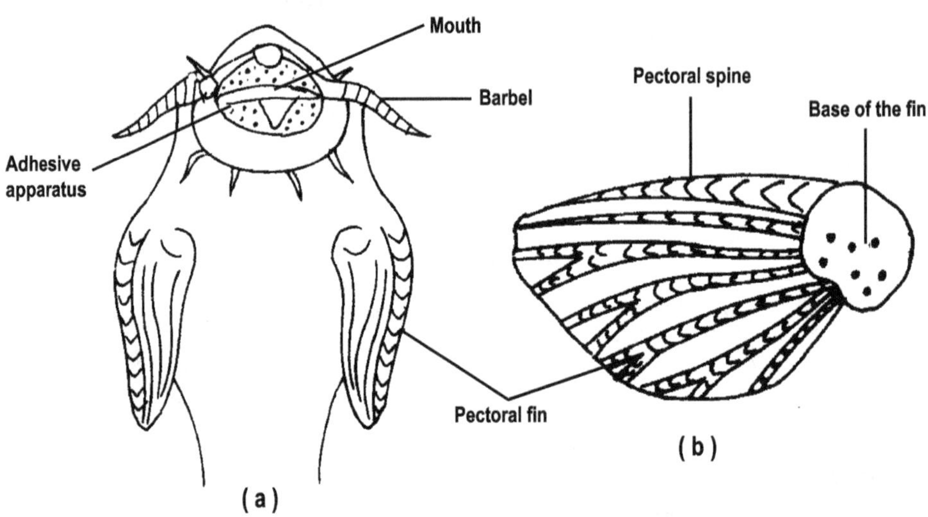

Figure 167: (a) *Glyptosternum* sp. Ahowing Adhesive Apparatus;
(b) Pectoral Fin of *Glyptosternum* sp.
(After Khanna and Singh, 2003)

The adhesive organs are inverted 'U' or 'V' shaped which is formed by the folds of skin between the bases of the pectoral fins *e.g., Glyptothorax* sp. and *Pseudoecheneis* sp. (*Figures 168, 169*).

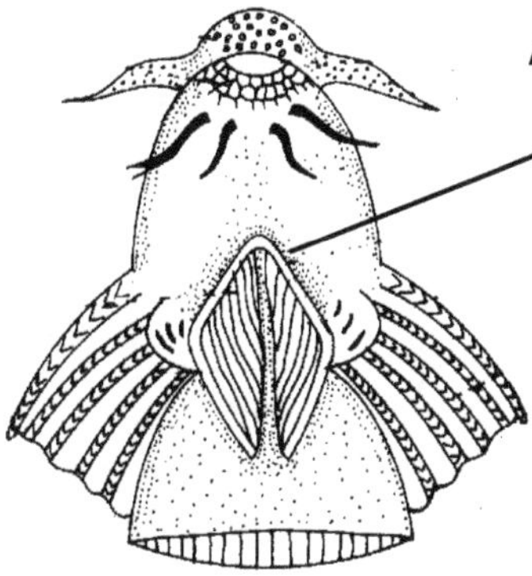

Figure 168: ***Glyptothorax pectinopterus*** **(Ventral Side) (After Khanna and Singh, 2003)**

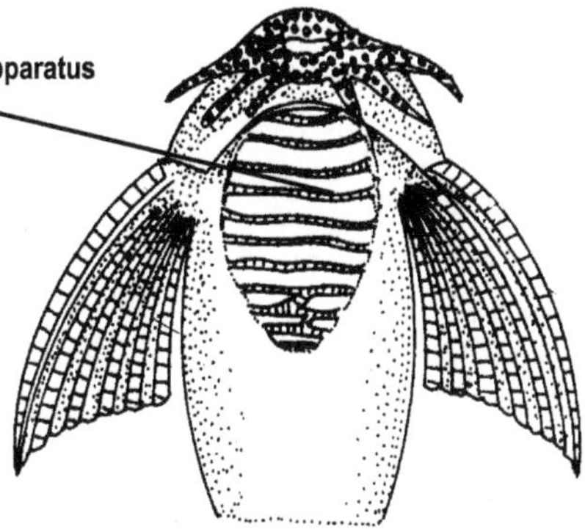

Figure 169: ***Pseudecheneis sulcatus*** **(Ventral Side) (After Khanna and Singh, 2003)**

Body Form

Shape

Commonly the fish body is torpedo-shaped (fusiform) and most often slightly to strongly ovoid in cross-section. However, in free living forms it becomes streamlined. Many fishes depart moderately to completely different from existing generalised shapes. These departures have been given the terminology as follows:

Symmetrical Body

Majority of the fishes have bilaterally symmetrical body [*Figure 170* (*a–h*)] but their shapes vary as:

Fusiform

The fish has a contour of a torpedo, *i.e.*, it is streamlined or boat-shaped. The contour gently sweeps back from a maximum girth in cross-section at about 1/3rd of the length from the anterior end (*Figure a*), *ex.* tuna.

Globiform

The profile of the body is globe-like *i.e.*, round in cross-section (*Figure b*), *ex.* puffers.

Compressed Form

The body profile is laterally compressed (*Figure c*), *ex.* butterfish.

Serpentine Form

The body is snake-like *i.e.*, cylindrical, elongate and sub-cylinderical in cross-section (*Figure d*), *ex.* eel.

Filiform

The body is thread-like *i.e.*, highly attenuated profile (*Figure e*).

Ribbon-like Form

The body is elongate but compressed laterally (*Figure f*), *ex.* ribbon fish.

Depressed Form

The body is flat dorsoventrally *(Figure g), ex.* catfish.

Asymmetrical Body

Some fish like flat fish are exceptional in having a bilaterally asymmetrical body *(Figure h).*

The body of the fish may be compressed laterally or depressed from back to abdomen.

Head Region

In this region the arrangements of various processes (eyes, mouth, snout, jaws, teeth, barbel etc.), their shapes and structures play an important role in identification.

Eyes

Eyes are subcutaneous and mostly placed dorso-laterally along the mid-axial line of the body. The variations range from reductions that produce blindness (cavefishes–amblyopsidae, hagfishes, myxinidae, benthobathis etc.) through those which enable vision in air as well as water (*Anabas* sp., *Periophthalmus* sp.).

Position

Their position may vary. Sometimes they are placed on the top of the head (superior) (*Figure 171*), or placed low (inferior) so that they are visible from the ventral side (*Figure 172*).

Size

Eyes are generally large or moderate in size but in eels and hill stream fishes they are small or in some cases even minute or completely hidden under the skin (*Figure 173*).

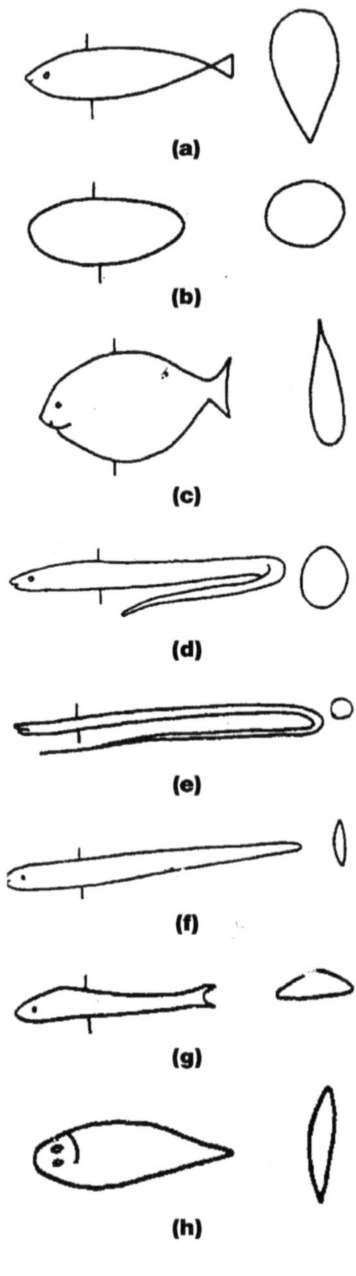

Figure 170(a–h): Body Forms

Shape

Eyes may be circular or oval in shape with circular or oval eyelids *(Figure 174).* Some fishes which dwell in caves are totally blind *(Figure 175).*

Eyes are generally without eye lids except for some which have nictitating membranes (sharks) and few bony fishes with fixed fatty (adipose) eyelids.

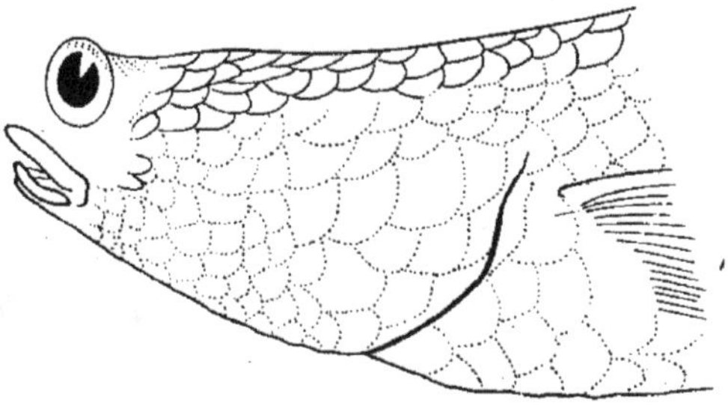

Figure 171: Eyes Superior (*Mugil corsula*)

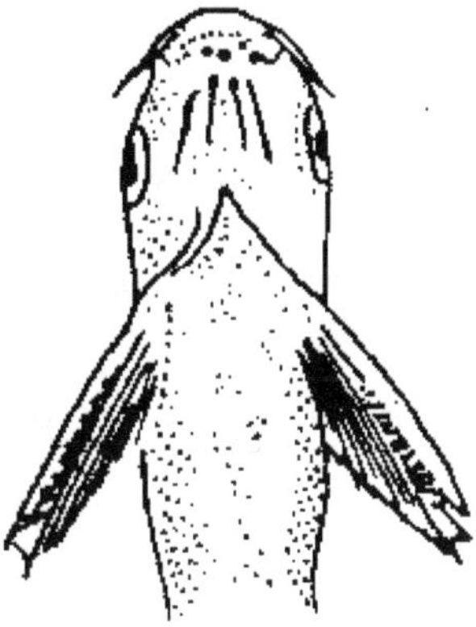

Figure 172: Eyes Inferior Visible from Below Ventral Surface (*Chandramara chandramara*)

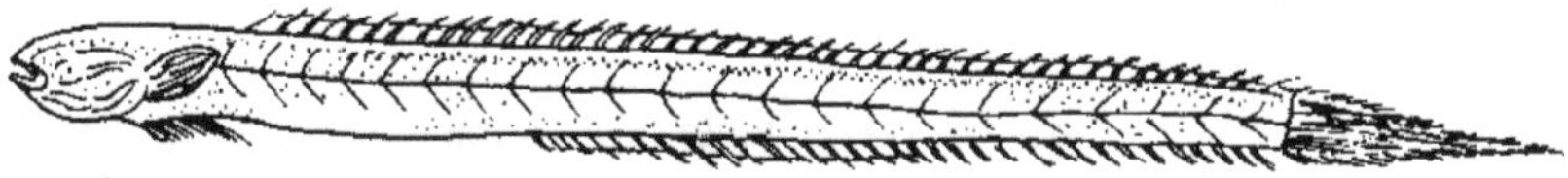

Figure 173: Eyes Minute, Reduced, Hidden (*Brachyamblyopus burmanicus*)

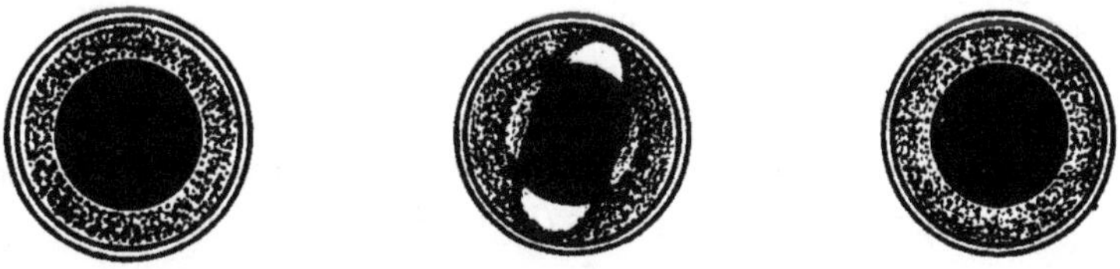

Figure 174: *(From left)* Eyes with Circular Eyelids; Eyes with Lateral Eye Lids; and With an Adipose Growth of Skin

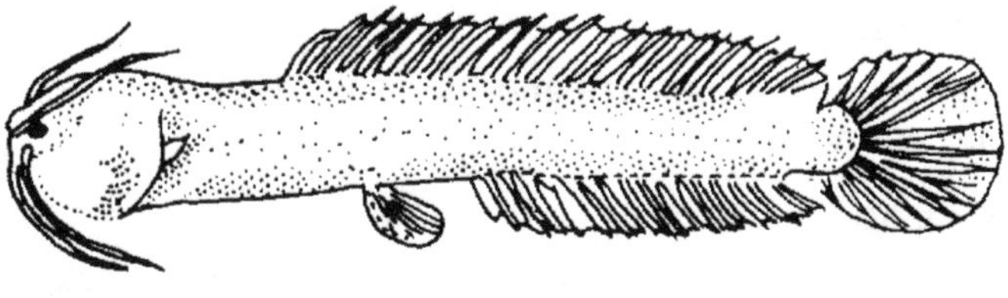

Figure 175: *Horaglanis krishnai* (A Blind Fish Living in Wells of Kerala)

Mouth

It is the chief feeding organ for fish. Its shape, position and structure varies depending upon the type of food it takes.

Position

The mouth in case of fishes may be [*Figure 176 (a–d)*]:

Dorsal *(Superior)*: When it is lying above the dorsal axis (*Figure a*).

Terminal: When it is placed on body axis (*Figure b*).

Ventral (Inferior): When it lies on the underside (*Figures c and d*) of the body.

Semi-ventral *(inferior)*.

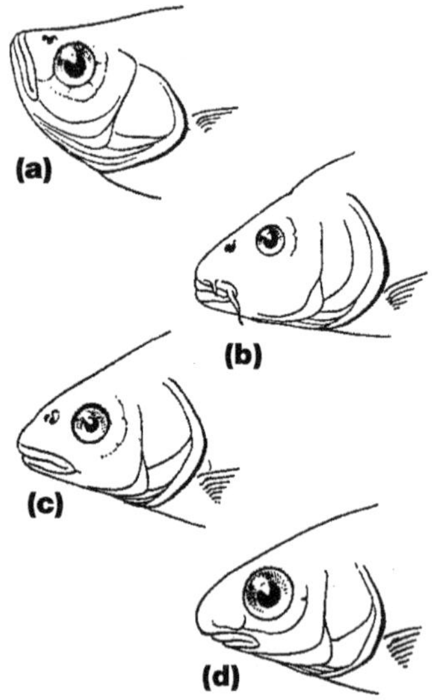

Figure 176 (a–d): Positions of Mouth

Shapes

Structurally mouth has been modified into various shapes *e.g.*, elongated-beaks, half-beaks, thick-lipped, star-gazers, paddle-like [*Figure 177 (a–g)*].

Structure

The mouth can be classified on the basis of its structure and function [(*Figure 178 (a–e)*]:

Grasping Mouth

Large with sharp teeth on both the jaw bones and often on the vomerine and palatine bones *e.g.*, pike, perch (*Figure a*).

Mouth Forming a Sucker

No jaws, teeth in the form of horny tubercles, a tooth present on tongue (basic boring organ) *e.g.*, lampreys and hag fishes (*Figure b*).

Imbibing Mouth

In the form of tube of variable length, sometimes extensible, usually toothless bottom feeders *e.g.* mormyridae, bream, pipe fish (*Figure c*).

Planktophagic Mouth

Usually large or of medium size and immobile. Teeth small or lacking *e.g.*, white fish, some cyprinoids etc. (*Figure d*).

Periphyton Eating Mouth

Placed on lower side of the head, shaped like a transverse crack, lower lip has sharp cutting edge. Sometimes with horny covering, teeth lacking *e.g.*, *Schizothorax* sp. etc. (*Figure e*).

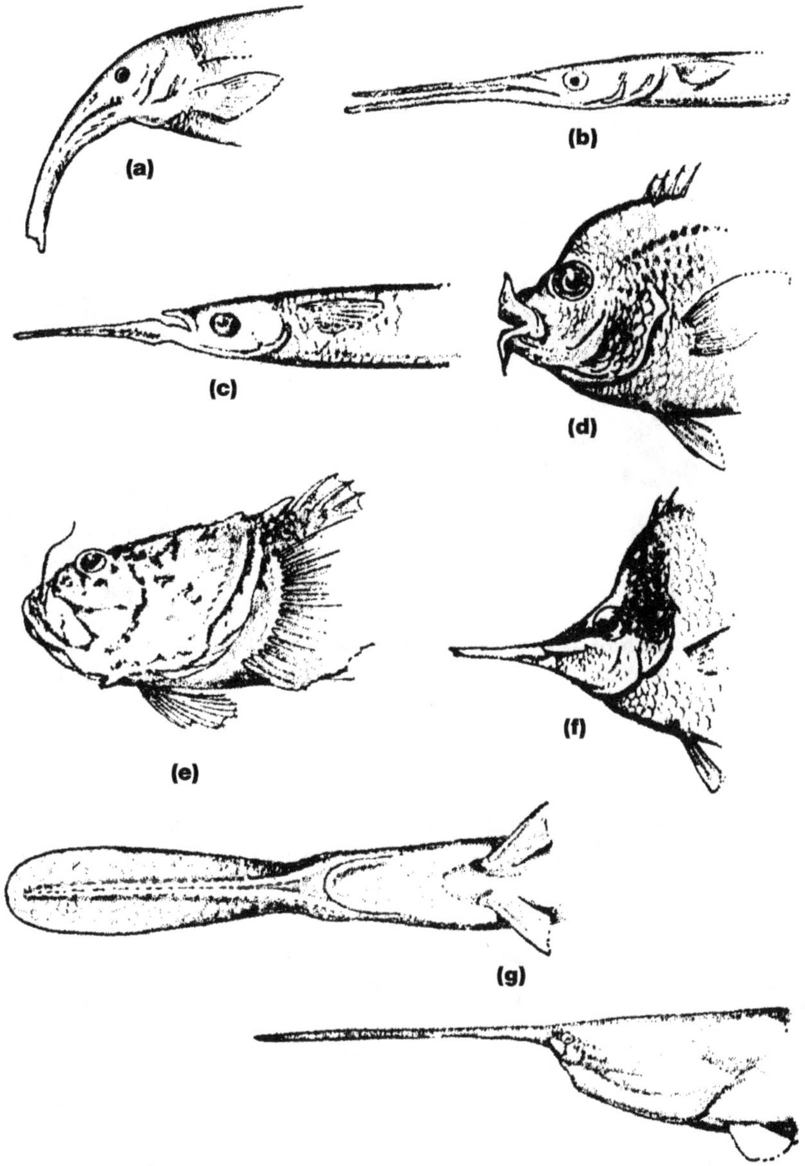

Figure 177: Shapes–(a) Elongated Beaks; (b) Elongated Beaks; (c) Half Beaks; (d) Thick Beaks; (e) Stargazers; (f) Long Nose; (g) Paddle Like (After Lagler *et al.*, 2003)

Crushing Mouth

Sometimes in the form of beak (trunk fish) with powerful teeth, sometimes plate-like (skates) or with spines; usually serves for crushing invertrebrates such as molluscs, echinoderms etc.

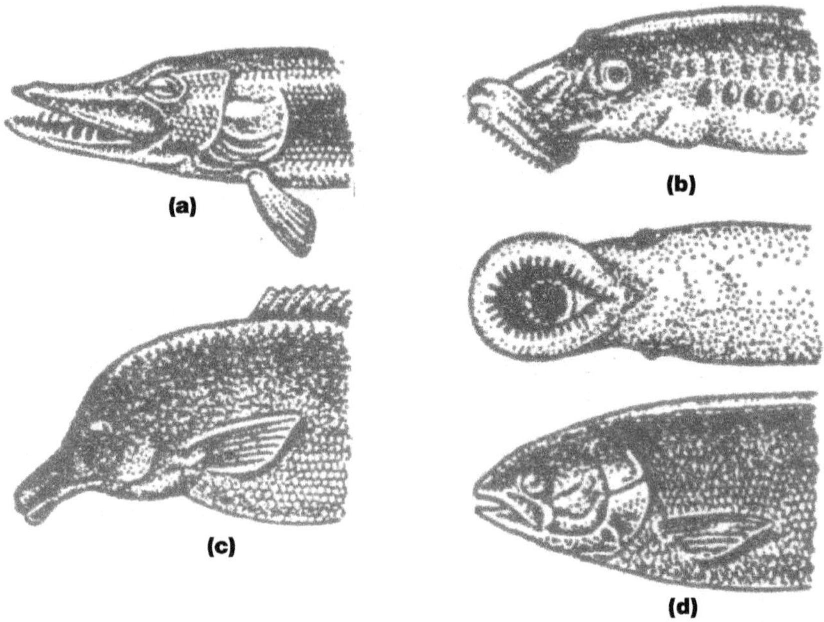

Figure 178: Structure–(a) Grasping Mouth (Pike); (b) Sucker Mouth (Lamprey); (c) Imbibing Mouth (*Mormyrus* sp.); (d) Planktophagic Mouth (*Salmon* sp.) (After Nikolsky, 1999)

Snout

Whole of the head infront of eyes is snout [*Figure 179 (a–d)*]. It may be smooth in most of the cases but in some, tubercles may be seen (*Figure 163*). Variations may be pointed (*Figure a*), rounded (*Figure b*), overhanging the mouth (*Figure c*) or tubular (*Figure d*).

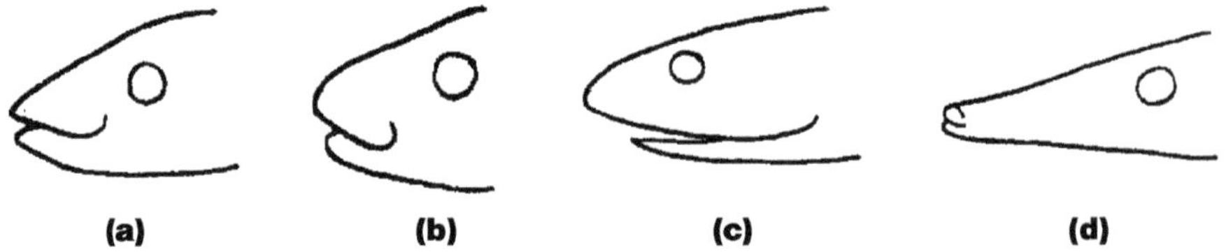

Figure 179: (a) Snout Tip Pointed; (b) Snout Tip Rounded; (c) Snout Overhanging Mouth; (d) Snout Tubular

Jaws

Jaws in most of the fishes are more or less of equal length but certain variations do occur [*Figure 180 (a–c)*]. Sometimes either both jaws are projecting (*Figure a*) or only one of the jaws (*Figure b*). In some lower jaws have a horny covering (*Figure c*).

Cleft

The cleft of mouth may or may not be oblique [*Figure 181 (a–b)*].

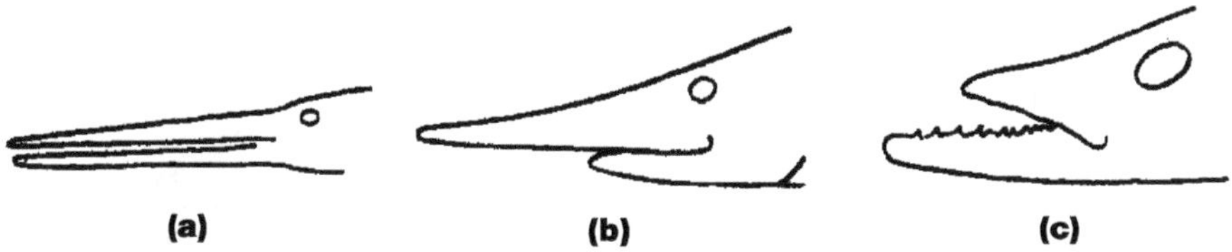

Figure 180: (a) Both Jaws Projecting; (b) Upper Jaw Projecting; (c) Lower Jaw Projecting (Teeth Canine)

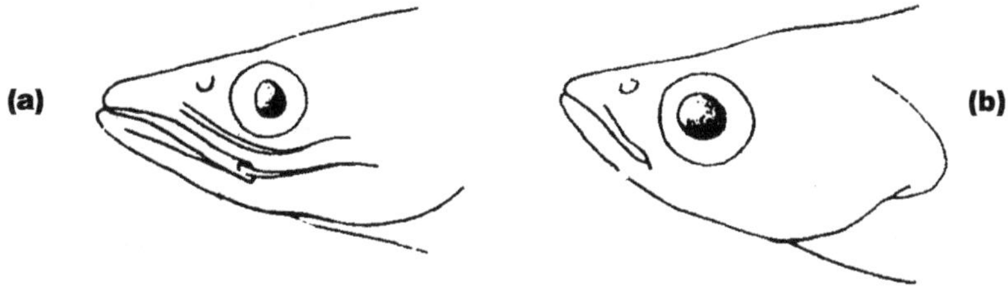

Figure 181: (a) Cleft Not Oblique; (b) Cleft Oblique

Lips

Lips border the mouth opening. Some fishes have unmodified thin lips which swallow large morsels of food, while the suctorial feeders have inferior mouth with fleshy modification of lips (*Figure 182*). These lips may be plicate (having folds) or papillose (having small tuft of skin or papillae). In parasitic members, the jawless suctorial mouth serves as a holdfast for attachment to the host and as a food remover from the host. In *Gyrinocheilus* sp. of southeastern Asia, the lips surround the mouth and form a funnel like sucker, serving not only to scoop up the mud, but also to enable the fish to cling to the stones and other objects (*Figure 183*). In mahaseers, the lower and upper lips are continuous around the jaws and the labial fold is uninterrupted by the isthmus (*Figure 184*) or interrupted (*Figure 185*). In *Tor progenius*, the upper lip is modified as a fan-shaped structure (*Figure 186*). Mostly these lips are thin smooth membranes but in some they may be with pores (*Figure 187*), stripes (*Figure 188*) or modified to form a sucker-like disc as in *Garra* species (*Figure 166*).

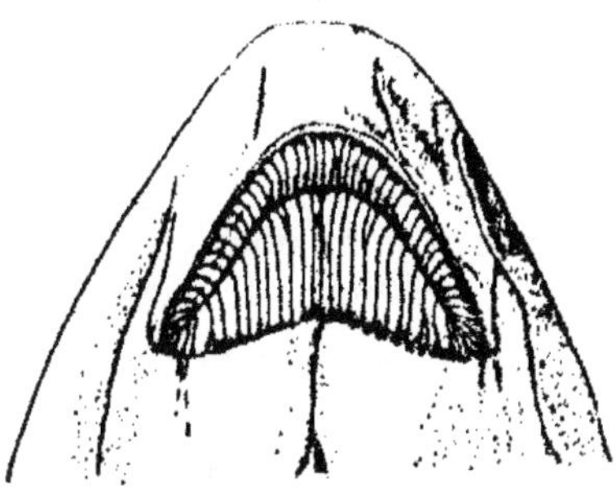

Figure 182: Sucker Mouth (*Moxostoma macrolepidotum*)

Figure 183: Lips Surrounding the Mouth (*Gyrinocheilus* sp.)

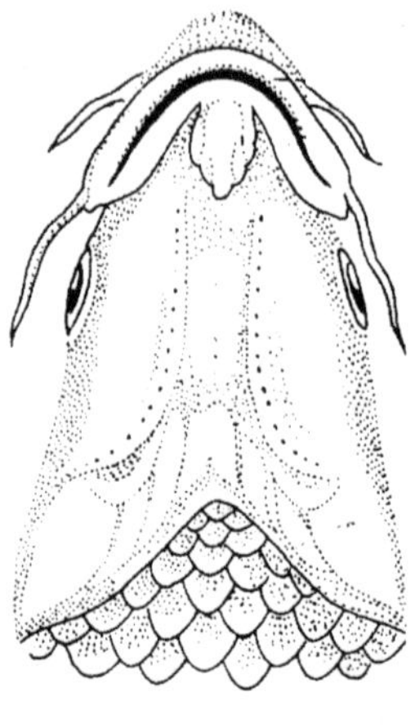

Figure 184: Labial Fold Uninterrupted by Isthumus

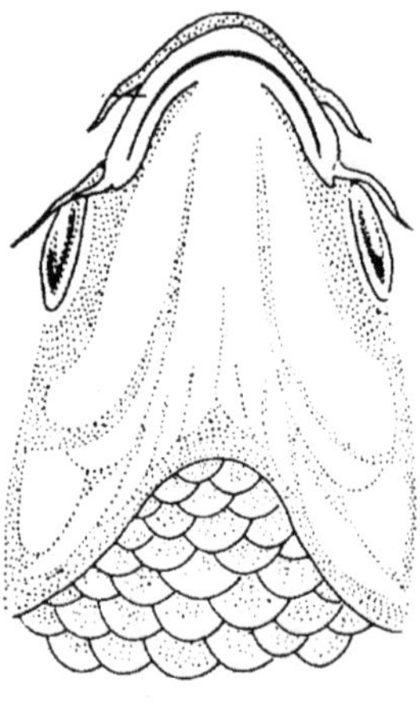

Figure 185: Labial Fold Interrupted by Isthumus

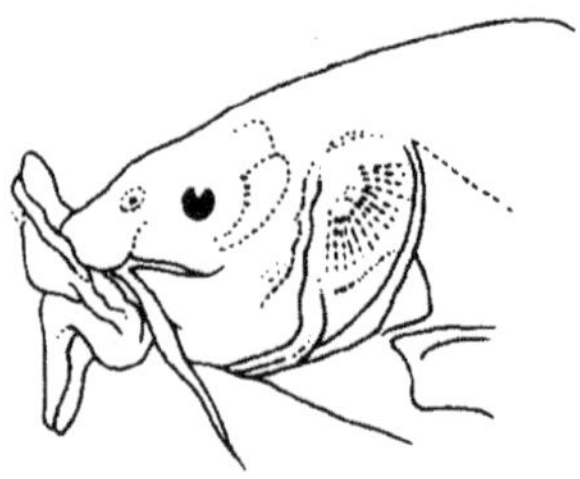

Figure 186: Fan-shaped upper lip (*Tor progenius*)

Figure 187: Lips with pores (*Labeo dero*)

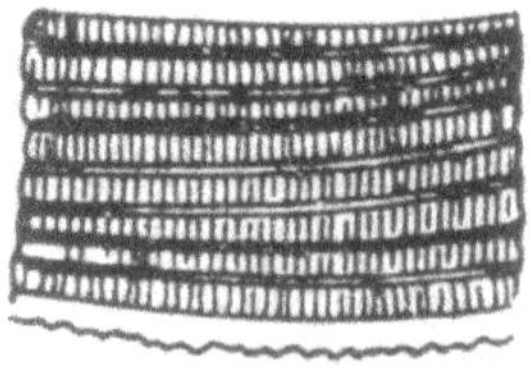

Figure 188: Lips with stripes (*Labeo dyocheilus*)

Teeth

Teeth arise from scales covering the lips [*Figure 189* (*a–e*)]. In general, teeth are absent in plankton feeders and in some omnivores. In bony fishes, teeth are of three types, based on where they are found–jaw, mouth and pharyngeal or rostrum.

1. Teeth when present may be:

   Cardiform: Numerous, short, fine and pointed resembling multi-toothed wool card of spinning wheel days (*Figure a*).

   Villiform: They are more or less elongated cardiform ones in which length to diameter relationship resembles that of intestinal villi (*Figure b*).

   Canine: These are dog-tooth-like or even fang-like elongated, sub-conical, straight or curved (*Figure c*).

   Incisors: Sharply edged cutting teeth. They may look like human, in others they are crenulated (saw edged) and in still others they are fused into cutting beaks (*Figure d*).

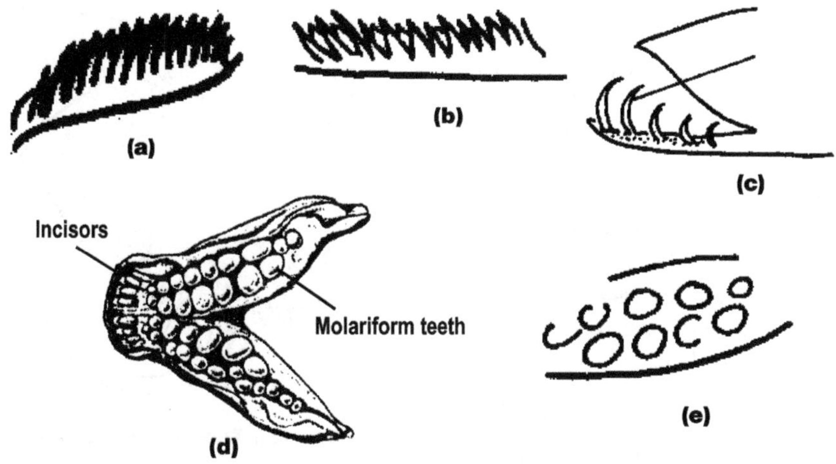

Figure 189: Teeth–(a) Cardiform; (b) Villiform; (c) Canine; (d) Incisors; (e) Molariform

Molariform teeth: They are for cutting and grinding and hence they have broad occlusal surfaces (*Figure e*).

2. Teeth may be found on sites other than jaws [*Figure 190 (a–b)*].

   Vomerine teeth: On vomer bone (*Figure a*).

   Palatine teeth: On palatine and ectopterygoid bones. In case of *Pristis* sp. the teeth are in the form of tooth-like denticles along the margin of saw-like rostrum (*Figure a*).

   Pharyngeal teeth: Occurs as pads on gill arches (*Figure b*).

   Tongue teeth: On the tongue.

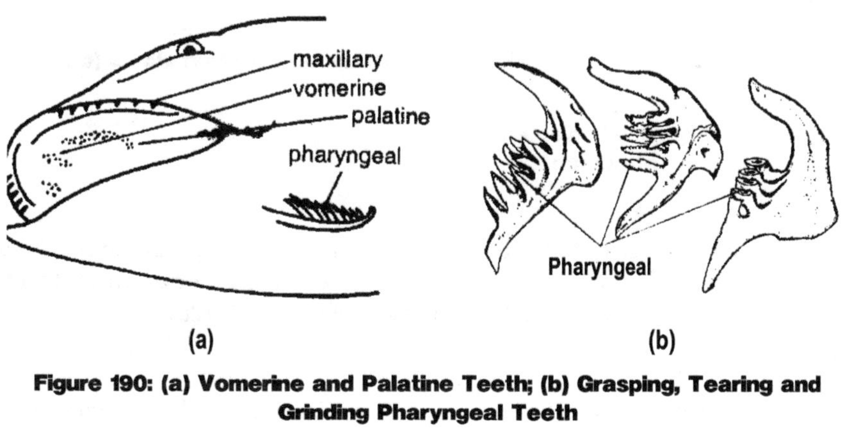

Figure 190: (a) Vomerine and Palatine Teeth; (b) Grasping, Tearing and Grinding Pharyngeal Teeth

Nostrils

These are a pair of apertures on the snout which is not communicated to the mouth. Nostrils are a vital part of fish and useful in classification. In lampreys and hagfishes there is a single median nostril (*Figure 3*). Most fishes have nasal opening at the top and dorsolateral side of the snout. However, in sharks, rays and skates, the noses are on the ventral surface of snout (*Figure 191*).

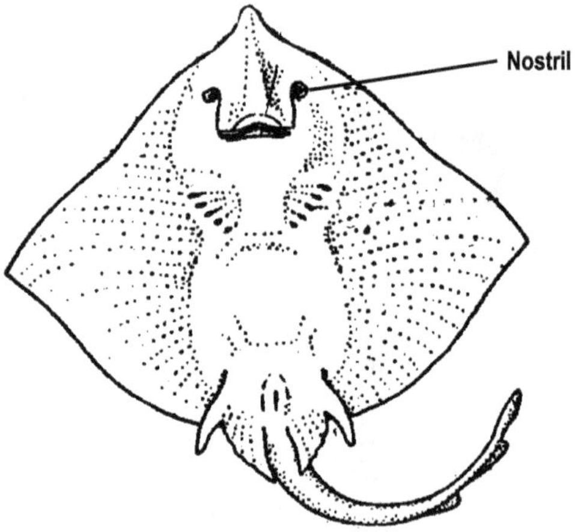

Figure 191: Thornback Ray (*Raia*

Trunk

The trunk in case of fishes can be distinguished as back, sides and abdomen. It generally passes into tail.

Abdomen in case of fishes is mostly rounded except in flat fishes, hill stream fishes and deep sea fishes. In some fishes it may be keeled (*Figure 192*) or it may have serrations (clupeids) *(Figure 193)* or may have adhesive apparatus (*Glyptothorax* sp.) (*Figure 168*).

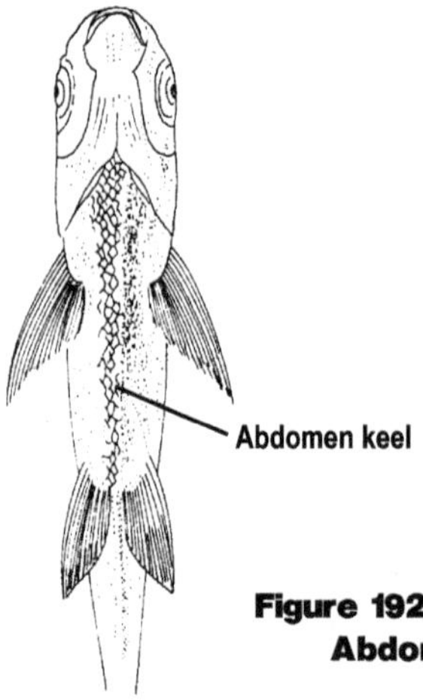

Figure 192: Keeled Abdomen

Figure 193: Abdomen with Serrations (clupeids)

The back and sides of the fish bear fins. In many fishes great part of the tail is surrounded by fins, leaving only a small portion (between dorsal, caudal and anal fins) finless and this part is called as peduncle of the tail (112a).

Tail

The portion behind anus is the tail and in fishes it has been represented by caudal fin and is of three main types:

Protocercal

It is considered as ancestral type that encloses vertebral column. It includes all types of symmetrical tails. Diphycercal tail is the secondarily symmetrical tail with equally developed lobes (*Figure 194*).

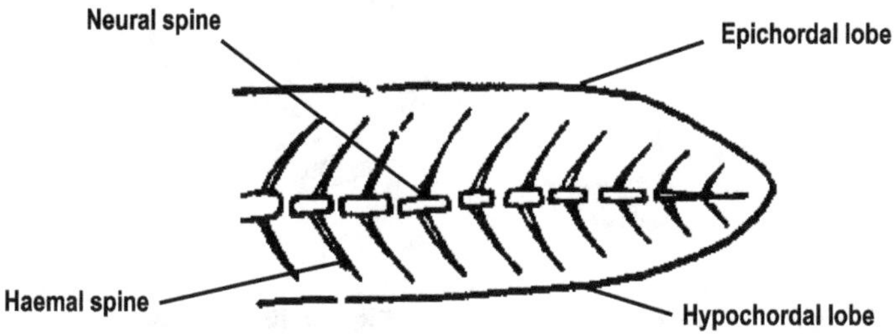

Figure 194: Protocercal Tail

Heterocercal

In this case the caudal fin is asymmetrical *(Figure 195)*. The vertebral column bents upwards or downwards and leads to asymmetrical lobes. It is further of two types:

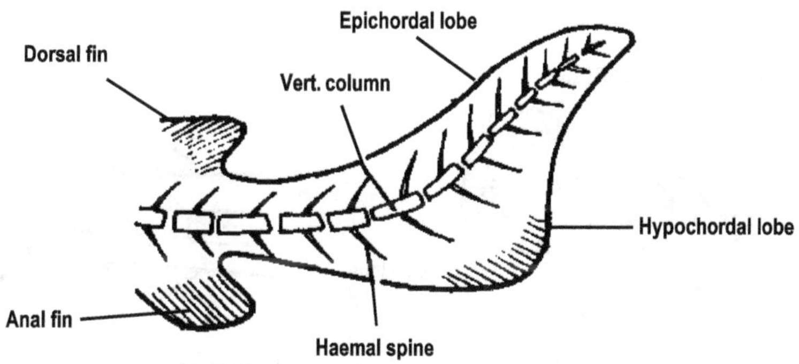

Figure 195: Heterocercal Tail

Epicercal

In this case the vertebral column bends towards the upper end and leads to the elongated upper lobe, whereas the lower lobe remains shorter (*Figure 196*).

Hypocercal

In this type the lower lobe is longer due to the extension of vertebral column, whereas the upper lobe remains shorter (*Figure 197*).

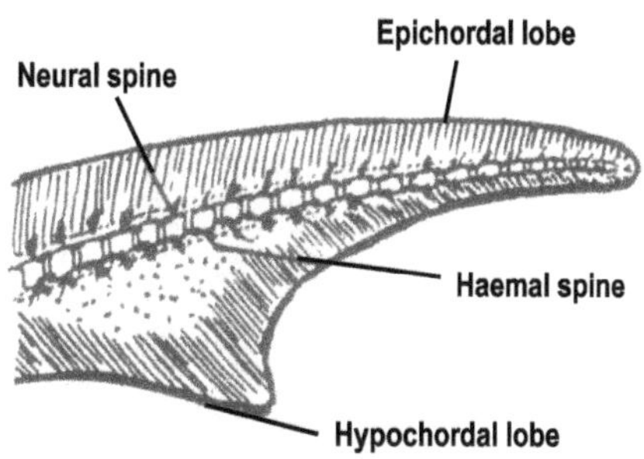

Figure 196: Epicercal Tail

Homocercal

The caudal fin in this case is symmetrical and consists of equal size of epi- and hypochordal lobe (*Figure 198*).

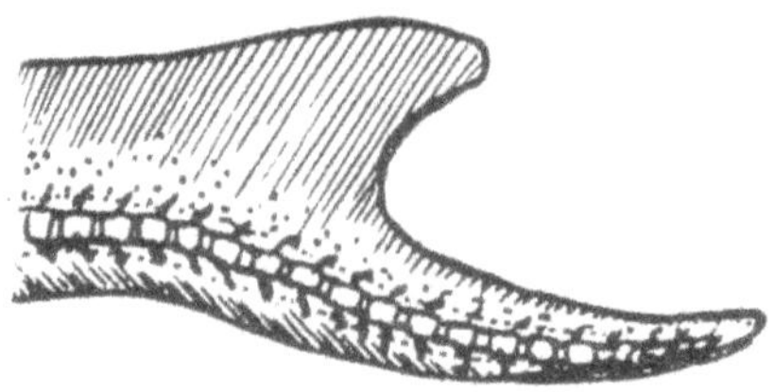

Figure 197: Hypocercal Tail

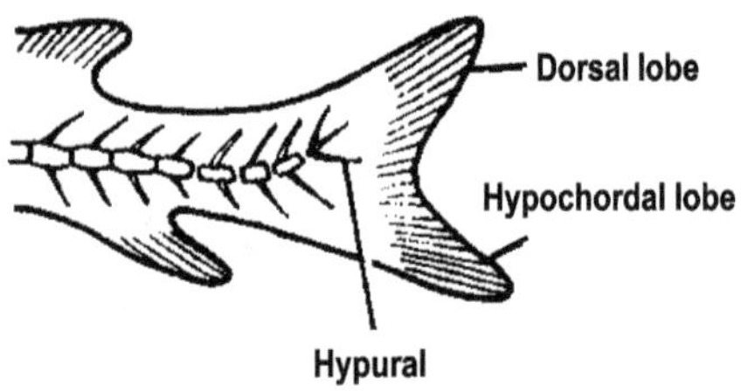

Figure 198: Homocercal Tail

Sometimes tail shows intermediate conditions between heterocercal and homocercal and is called as abbreviated heterocercal type. Here hypochordal lobe is greatly enlarged while epichordal lobe is reduced to mere vestige (*Figure 199*).

Sometimes there is a complete disappearance of caudal fin and a continuous median fin is formed by union of dorsal and caudal fin. Such tail is known as gephyrocercal tail (*Figure 200*).

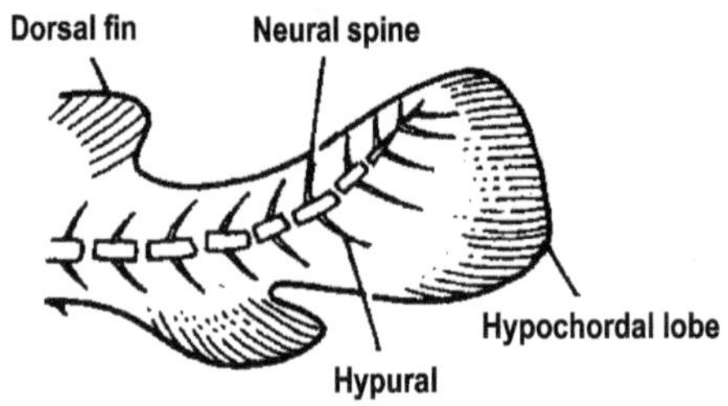

Figure 199: Abbreviated Homocercal Tail

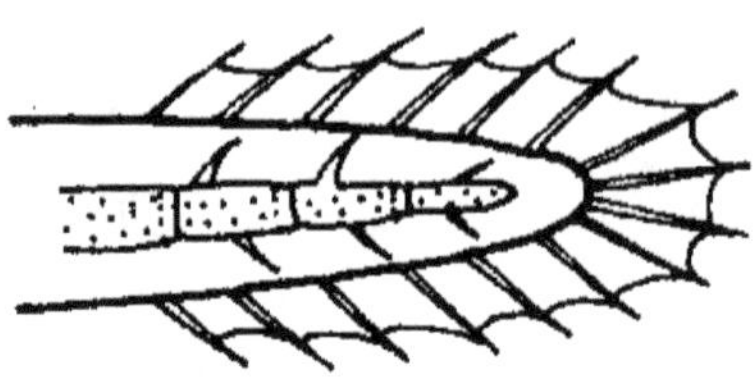

Figure 200: Gephyrocercal Tail

//•••••••••//

Chapter 8

Fins and their Modifications

Fins are the limbs of fishes and serve as a balancing organ during swimming. They vary in shape, size and position and are distinguished as unpaired and paired fins. The unpaired fins or median fins are dorsal (one or two) on the back, an anal on the lower edge of body behind vent and a caudal fin at the end of the tail (*Figure 201*). These fins are also called as marginal fins because they arise from the margins of body. The pectoral and pelvics are the paired fins and correspond to the fore and hind limbs of terrestrial vertebrates.

Fins are supported by skeletal rods called radials and fin rays. The flexible web of each fin is a double sheet supported by stiffy rods like spokes of a fan. The whole

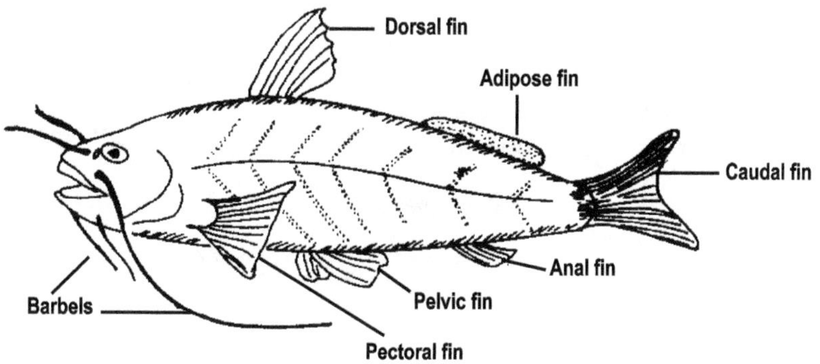

Figure 201: Unpaired and Paired Fins of a Fish

complex of a web, rods, internal supports and muscles which make up a fin is usually inserted into the body as a largely self-contained unit.

In teleosts, rays and spines arise embryologically as thickening in the basement membrane of the dermal fin fold. The opposed pairs become fused along most of the length to form shaft of the ray or spine. The halves remain separated at base where they are expanded to form a pair of tongs. As the fin rays approach the outer edge of the web, they branch force-and-aft several times and their shafts are divided into segments by regularly spaced rings. These rings provide flexibility to the rays.

In contrast to this, fin spines come to a point at their tips, rather than branching. The halves of shaft are at the base to form rigid and unsegmented spear.

In adult teleosts, spines, rays and internal supports are ossified; the bony rays are called lepidotrichia (homologous with scales). In Elasmobranchs, the fins are supported by ceratotrichia (composed of collagen) and they are more closely packed than teleost lepidotrichia but are also paired structures, each grasping a median cartilagenous support. Actinotrichia, the very fine horny filaments occur in fin fold of larval teleosts and may persist in outer edge of adult fin membrane.

The dermal fibres (ceratotrichia) are homologous with fibrous rays (actinotrichia), which are localised at distal end of bony fins in teleosts and also within adipose fin of salmonids, siluroids, characins and certain other fish.

Most cat fishes have large spines in their dorsal and pectoral fins. Such fins also occur in variety of other fishes and the function of this spine is protective.

Structure of Fin Rays

In teleosts, the fin rays are branched and consists of a number of lepidotrichial segments. Each segment consists of a pair of plate like elements called hemisegments separated by a thin band of tissue. The area inbetween two hemisegments is known as intrasegmental region and between adjacent segments is known as intersegmental region.

Each median fin is supported by radials (bony rods) and dermal fin rays (dermotrichia) (*Figure 202*). A radial consists of three pieces, of which proximal is

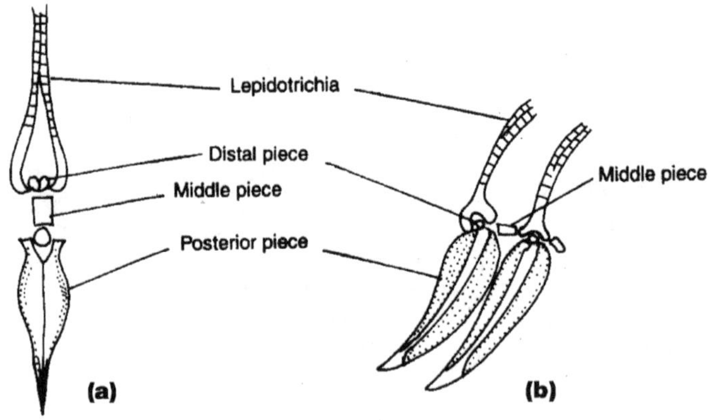

Figure 202: Radial and Dermal Rays–(a) Front View; (b) Lateral View

large dagger-shaped and is called ptergiophore and lies between the neural spine. There are two chief kinds of dermotrichia, the unbranched and unjoined horny fin rays or ceretotrichie of chondrichthyes and bony lepidotrichia of teleosts, usually branching and joined. Delicate horny rays (actinotrichia) develop in embryonic fins of all teleostomes and are homologous with ceratotrichia. They persist at growing margin of fins and are functionally replaced in later stages by lepidotrichia formed from modified scales (*Figure 203*). The dermal rays are of three kinds in teleosts.

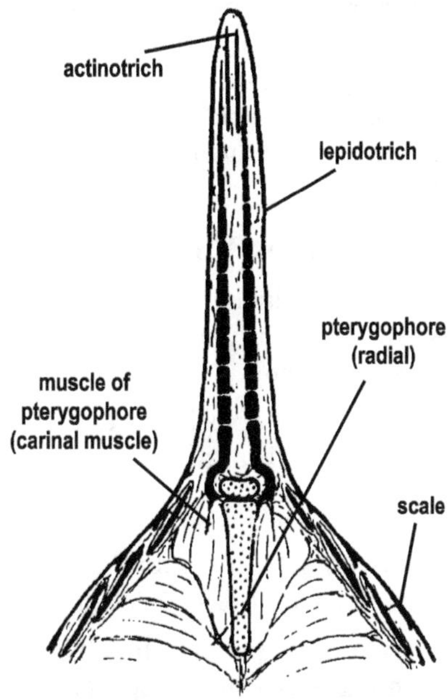

Figure 203: Schematized Section Through the Dorsal Fin of a Teleost (After Goodrich, 1984)

Fin Spine or Actinotrichia

These are hard unsegmented structures often found in the first dorsal fin of certain species. They develop as horny rays in all teleosts (*Figure 204*).

Soft Rays

These are flexible, segmented and branched structures. These are called as lepidotrichia and are arranged biserially in several species of teleosts. During development, they replace the actinotrichia except at the tip of the fin and are believed to be modified dermal scale (*Figure 206*).

Spinous Rays

In some fishes (carps and cat fishes), some soft rays appear to fuse together to form hard structures like spines (*Figure 205*).

Figure 204: Spine

Figure 205: Spinous Tay

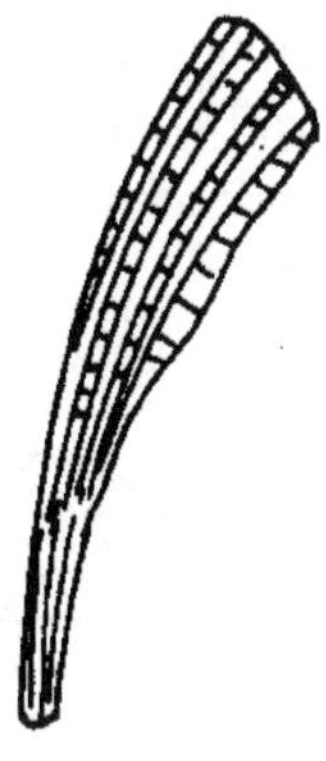

Figure 206: Soft Ray

Fins perform important functions in swimming and are of great help in starting, turning, steering and balancing the body in water. These functions are achieved by folding, opening or bending the fin blade which is supported by rays and spines. The term camptotrichia has been proposed for the rays of fins in the lung fishes (dipneusti) that have been thought to differ microscopically from those of all other fish organs.

Modifications/Variations

Variations in the skeleton have resulted in differences in both median and paired fins of fishes, many of which appear to be of adaptive significance for locomotion and maneuvering.

Different shapes in the dorsal, anal, pectoral and pelvic fins have been shown in the *Figures 209* and *210.*

Dorsal Fin

The dorsal fin exhibits a great diversity in size as well as forms and these modifictions are performed for a peculiar functions. Some of these are as follows:

The dorsal fin may be absent as in sting ray and eagle rays (*Figure 207*) or reduced as in *Notopterus* sp. (*Figure 207*).

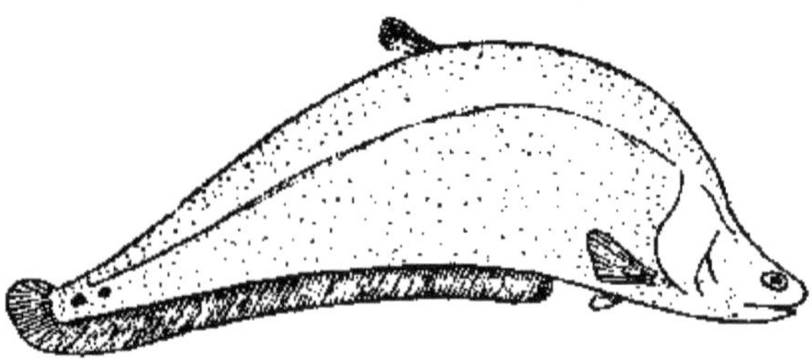

Figure 207 : *Notopterus chitala* (rudimentary dorsal fin)

In some of the sharks (heterodontidae) each of the two dorsal fins is preceded by a stout sharp spine (*Figure 208*). This spine is used for defensive purpose.

The fin may extend through most of the length of the back as in the case of bow fin (*Amia calva*) (*Figure 211*) or extends throughout the length (*Figure 212*) or it may

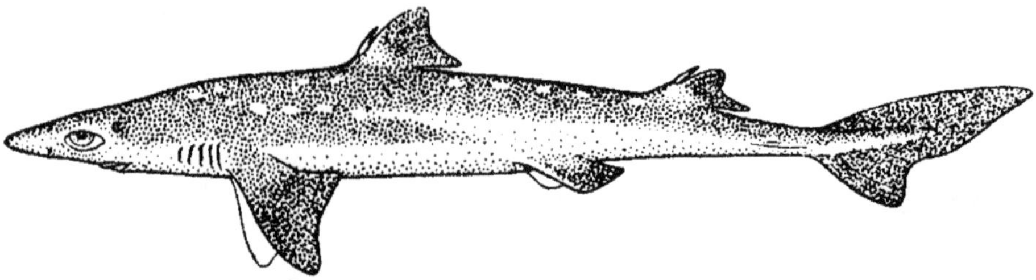

Figure 208: *Squalus acanthias*

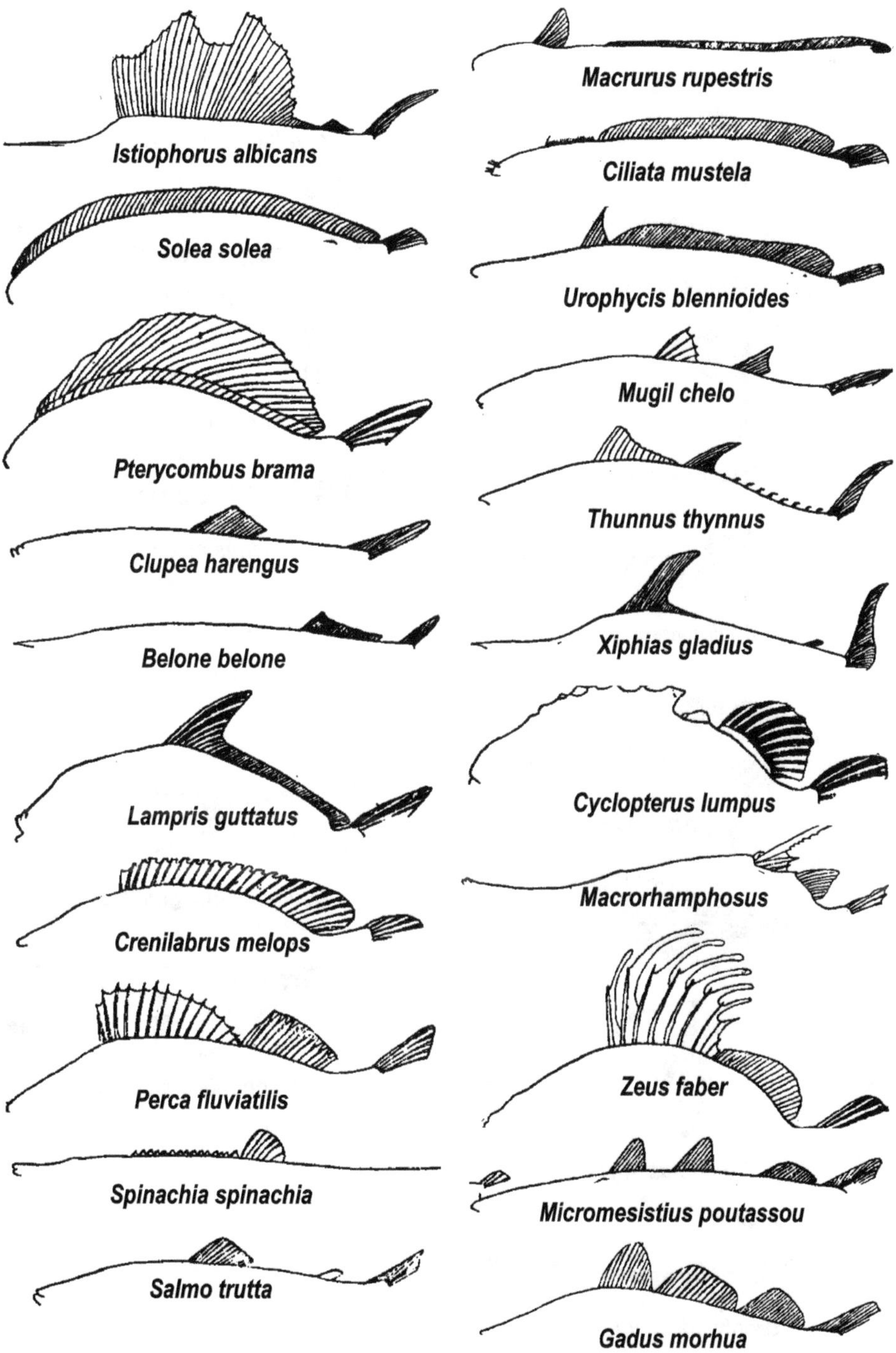

Figure 209: Shapes of Dorsal Fins (After Harder, 1975)

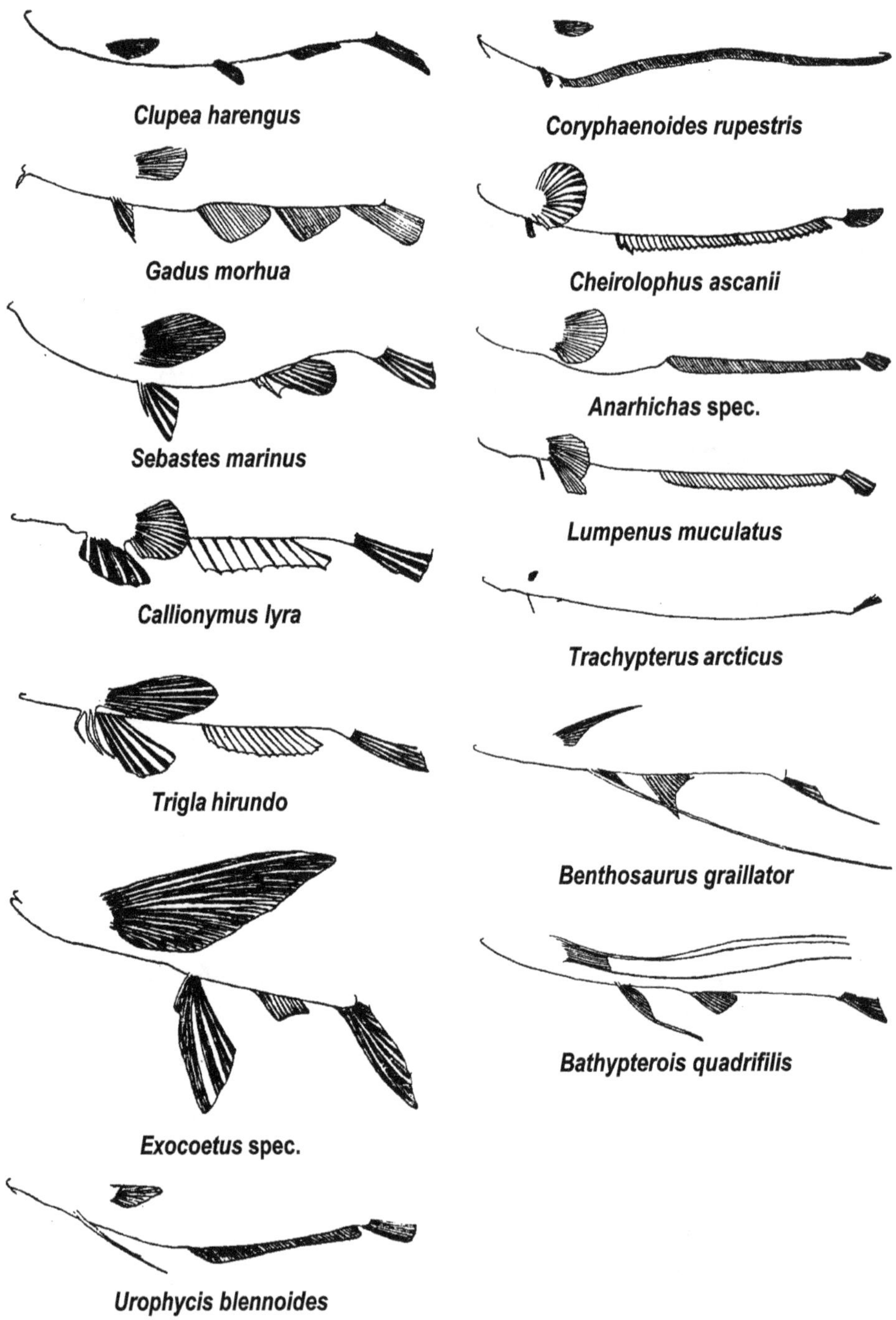

Figure 210: Shapes of Anal, Pelvic and Pectoral Fins (After Harder, 1975)

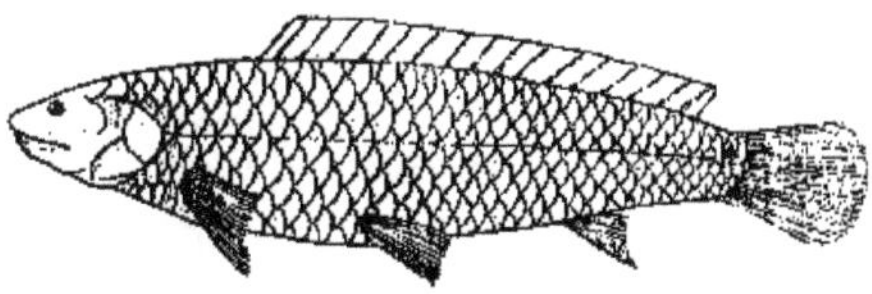

Figure 211: ***Amia calva***

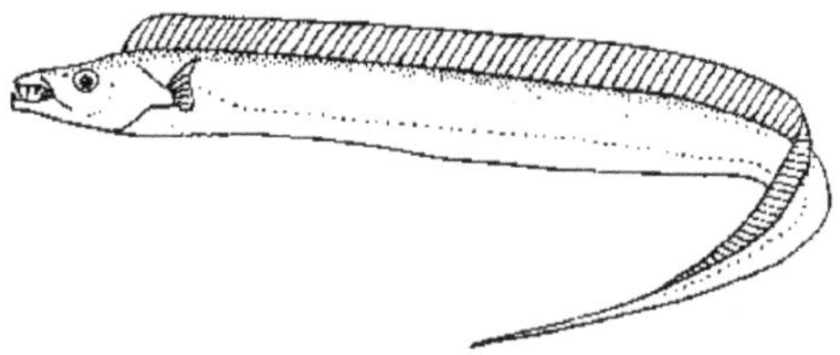

Figure 212: ***Trichiurus*** **(ribbon fish)**

be continuous with the caudal and anal ones as in muraenas and lampreys (*Figure 213*).

The dorsal fin may be two, three or even more in number as in the members of family percidae (*Figure 214*) and cods (*Figure 215*).

In large oceanic oar fish (*Regalecus glesne*) that attains length more than twenty feet, the fin extends all the way along the upper edge of compressed body and the first few rays are prolonged into rather long filaments, each of which ends in a membranous flap (*Figure 216*).

Figure 213: ***Petromyzon marinus*** **(Lampreys)**

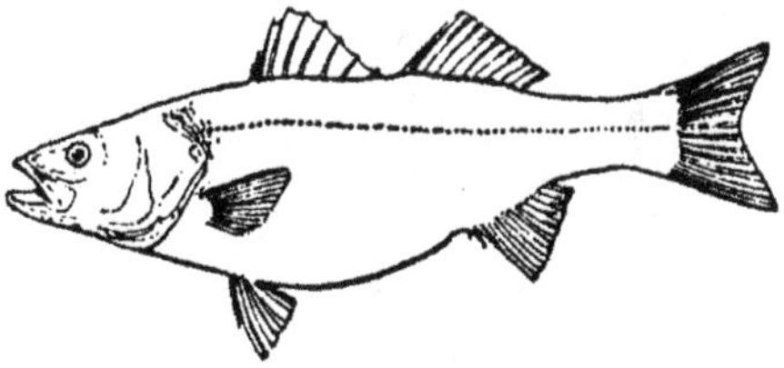

Figure 214: ***Roccus sexatilis*** **(Two Dorsal Fins in Percid)**

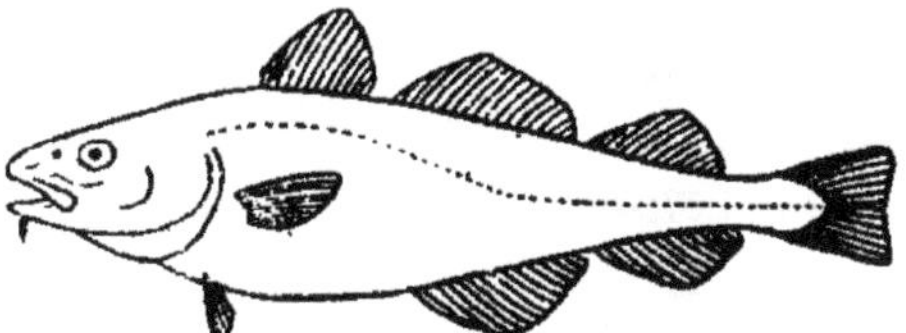

Figure 215: ***Gadus callarias*** **(Three Dorsal Fins in a Cod)**

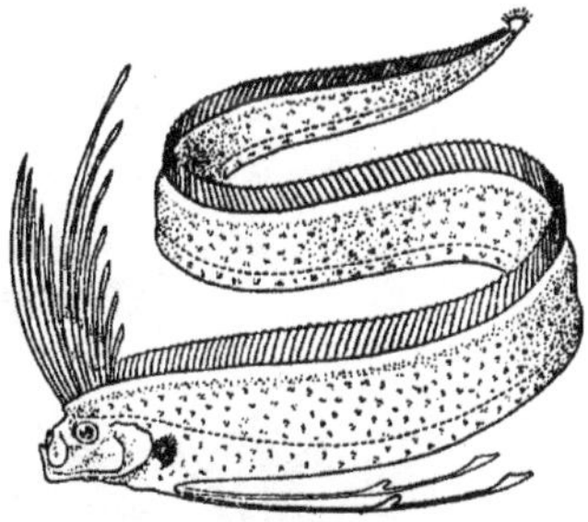

Figure 216: ***Regalecus glesne***

In young Deal fish (*Trachypterus arcticus*) the first six dorsal fin rays are produced into fine filaments more than four times as long as the fish itself (*Figure 217*). Another member of the same tribe, *Velifer hypselopterus* derives its name from the relatively huge size of both dorsal and anal fins (*Figure 218*).

In some fishes last rays of dorsal fin is elongated (*Figure 219*). In *Sillaginopsis*, second dorsal spine is prolonged as a long filament (*Figure 220*). In some fishes, the dorsal fin reduces itself into a series of more or less isolated finlets as in sticklebacks

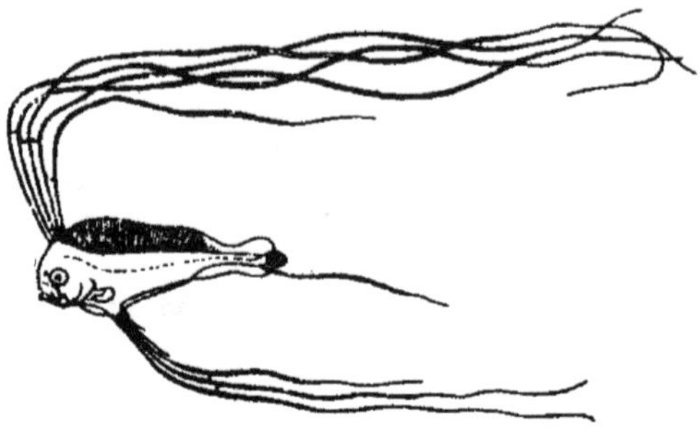

Figure 217: ***Trachypterus arcticus*** **(young)**

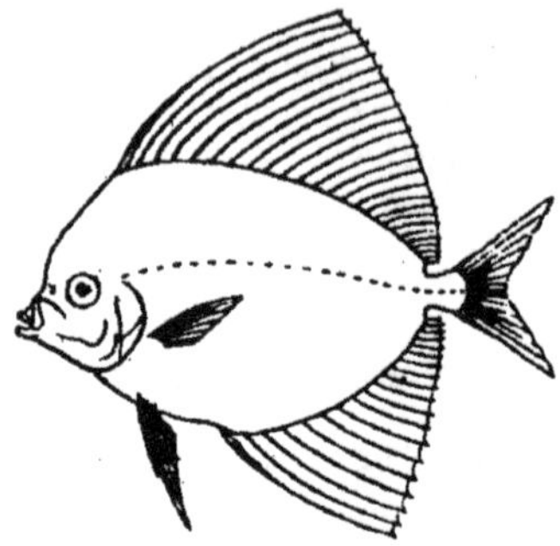

Figure 218: ***Velifer hypselopterus***

(*Figure 118*) and reed fishes (*Figure 221*). In mackerel (scomber), tunny (thunnus) and allied forms the soft dorsal fin is followed by a row of separate finlets (*Figures 222, 223*).

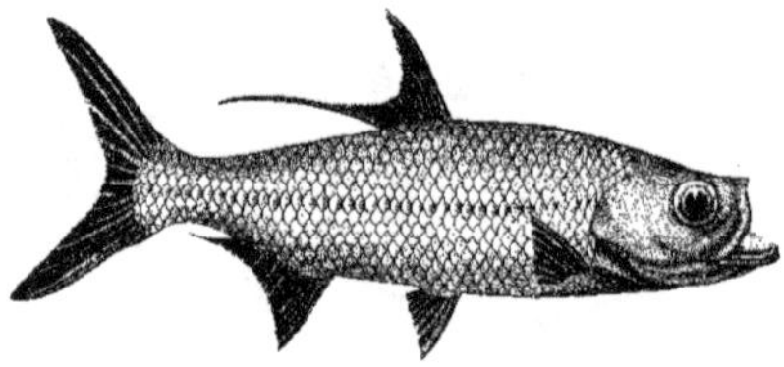

Figure 219: ***Megalops cyprinoides***

Figure 220: ***Sillaginopsis*** **sp.**

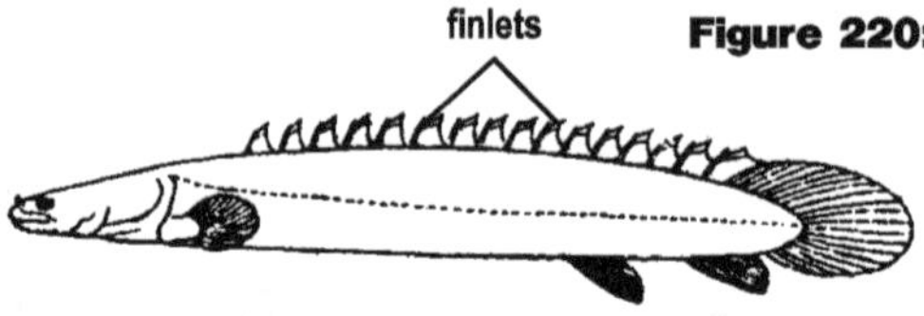

Figure 221: ***Polypterus bichir***
(Isolated Finlets in Reed Fish)

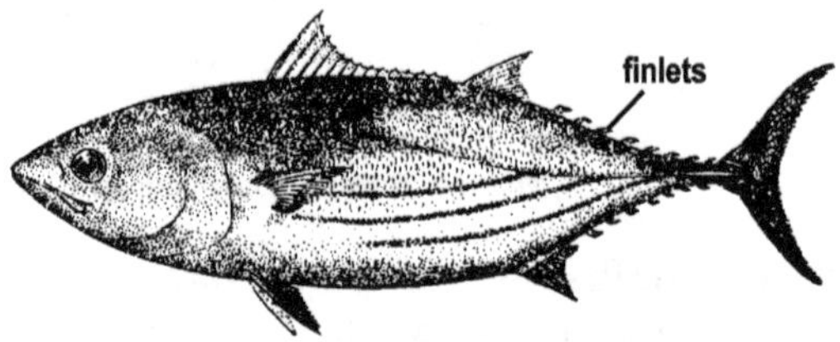

Figure 222: ***Gymnosarda pelamis***

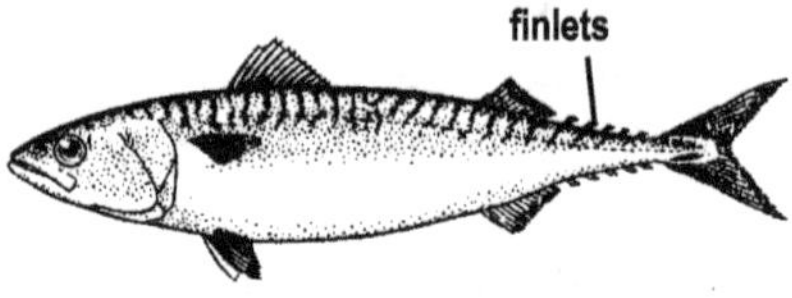

Figure 223: ***Scomber scombrus***

In Mastacembelus, the dorsal fin is in two parts; the first one with 32-40 short depressible spines and 46-90 rays (*Figure 224*).

Some fishes like trouts and their relatives (salmon family, many characins and majority of cat fishes) possess two dorsal fins, one of which possess no rays at all

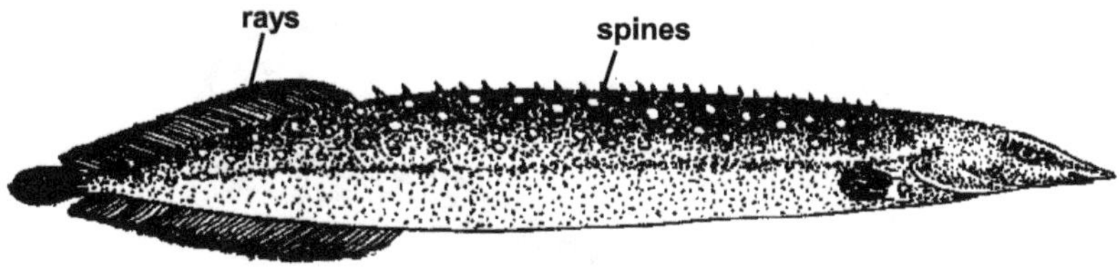

Figure 224: ***Mastacembelus armatus***

and is a fleshy structure and covered with skin. This is known as adipose fin (*Figures 20, 225).*

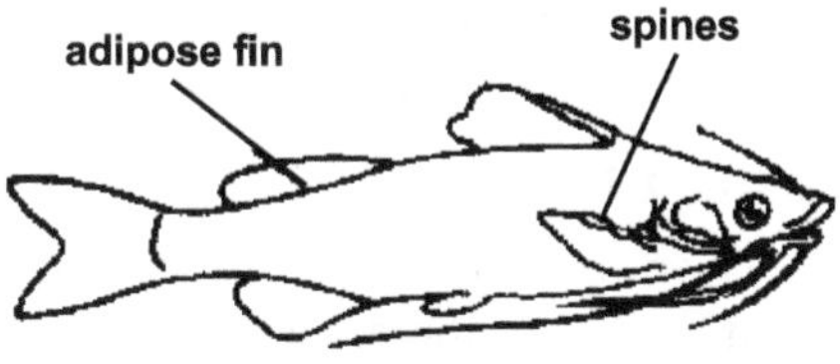

Figure 225: Catfish

The dorsal fin in most fishes varies in shape. It may be concave with smooth spine (*Figure 226*) or with serrated spine (*Figure 227*).

In some fishes dorsal fin may have a procumbent spine (*Figure 228*).

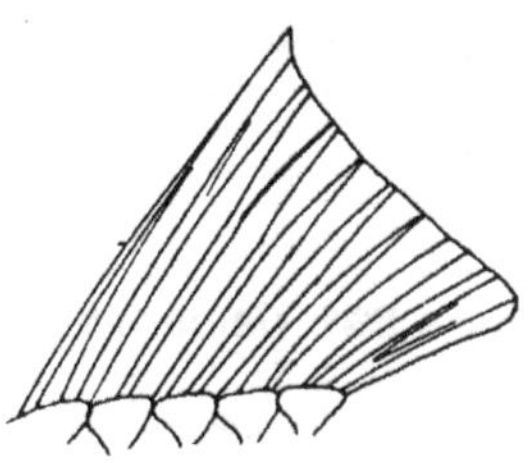

Figure 226: Concave with Smooth Spine

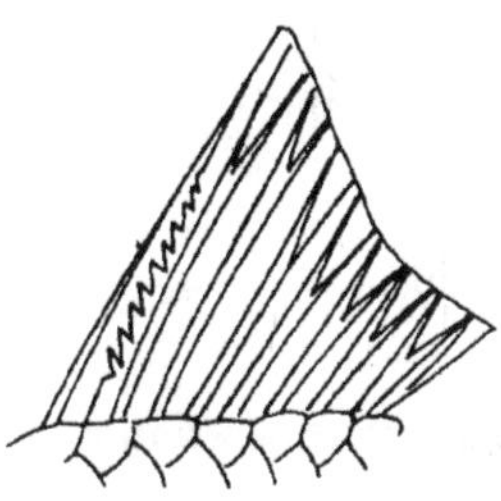

Figure 227: Concave with Serrated Spine

Figure 228: ***Mystacoleneus argenteus***

In swamp eels (symbranchids) the dorsal fin is vestigial and is in the form of fold or ridges only (*Figure 229*).

Dorsal fin may even have following unusual modifications.

Rod and Lure

In angler fishes (*Figure 230*), the first ray of dorsal fin is placed on the snout and is modified to form a fishing organ on head, the element for the pre-enticing "rod

Figure 229: ***Monopterus cuchia*** **(symbranchids)**

and lure" that adorns the top of the head and consists of a rod called the illicium and a bait called esca.

Sucking Disc

The dorsal fin in case of sucker fish (Remora) gets modified into a large sucking disc at the top of the head (*Figure 231*). This disc has 10-28 transverse ridges which are modified spines with which it attaches itself to the other fishes.

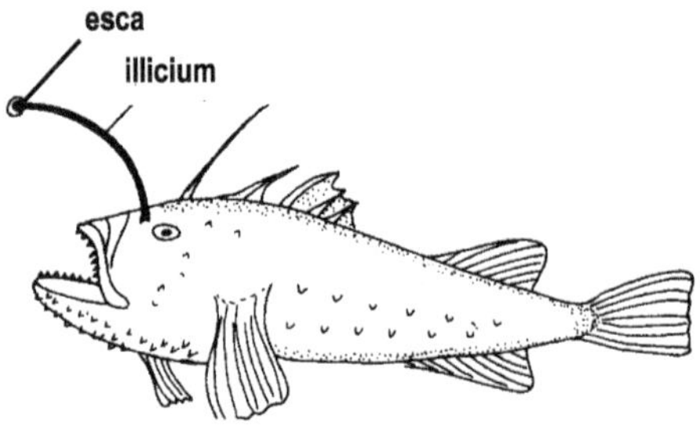

Figure 230: ***Antennarius*** **sp.**

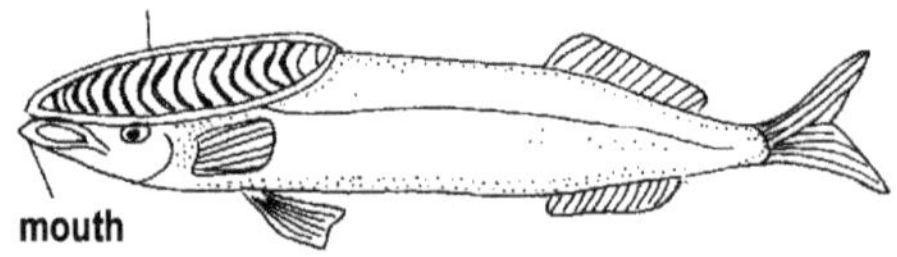

Figure 231: ***Remora remora***

The dorsal fin is wing-like in angel fish (*Figure 232*) or bizarre-shaped in scorpion fish (*Figure 233*) or highly extended (*Figure 234*).

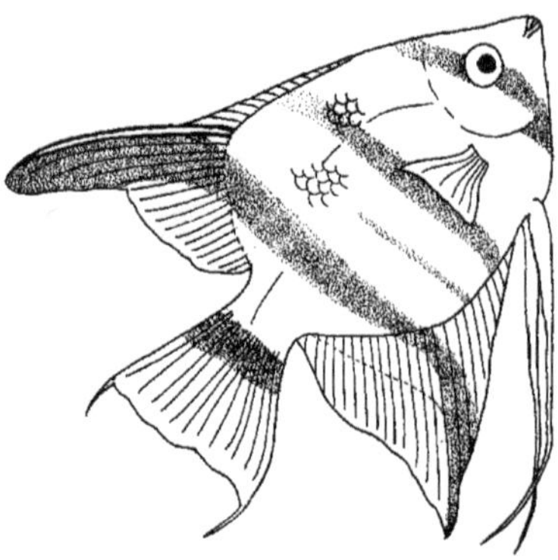

Figure 232: ***Pterophyllum scelare***

Figure 233: ***Pterosis volitans***

Position

The position of dorsal fin or fins exhibits variation in different fishes. In gar pikes, *Lepidosteus* species and the pike (Esox) both dorsal and anal fins are placed well back towards the hinder end of the fish (*Figure 13*).

In herrings (clupea) and carp (*Cyprinus carpio*) the dorsal occupies a position more or less in the middle of the back (*Figure 235*).

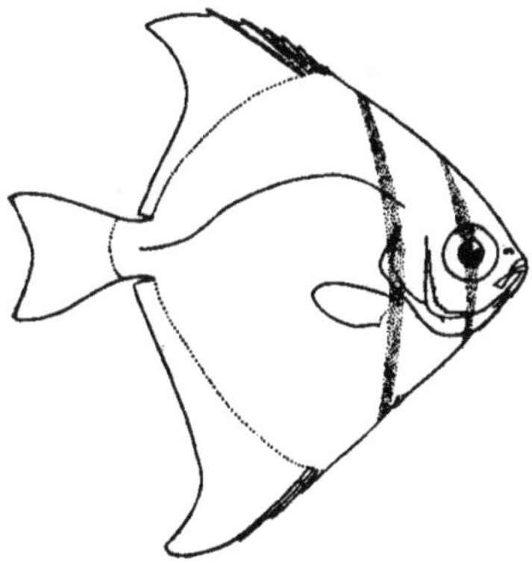

Figure 234: ***Monodactylus argenteus***

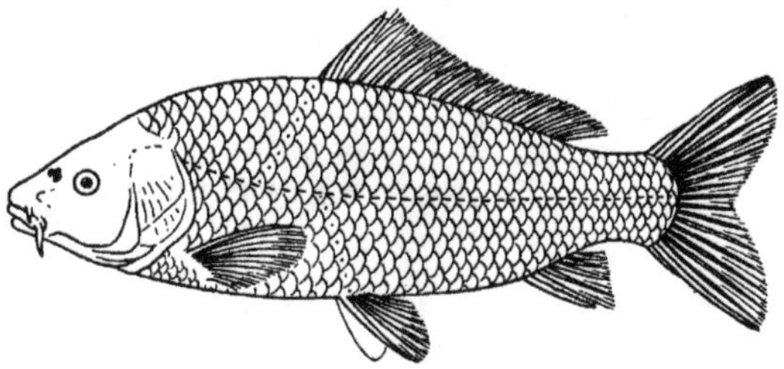

Figure 235: ***Cyprinus carpio***

In some cat fishes, the fin is considerably nearer to the head than to the tail *(Figure 236)*.

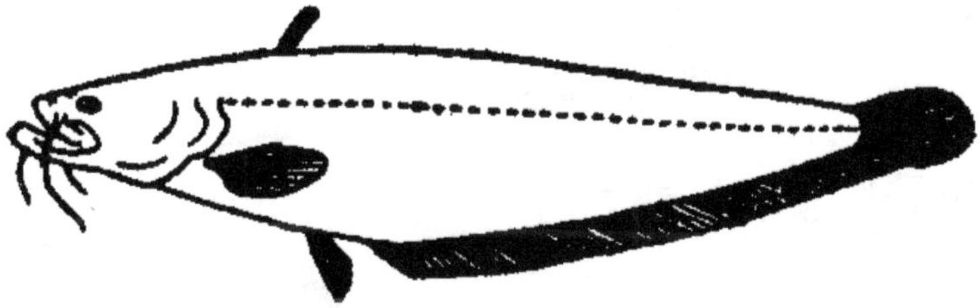

Figure 236: ***Silurus glanis***

The rays at the anterior end of the dorsal fin may also be transformed into spines as in seaperch *(Figure 237)* or the spinous position may be separated off as a distinct fin as in mackerel *(Figure 223)* or cardinal fish *(Figure 238)*.

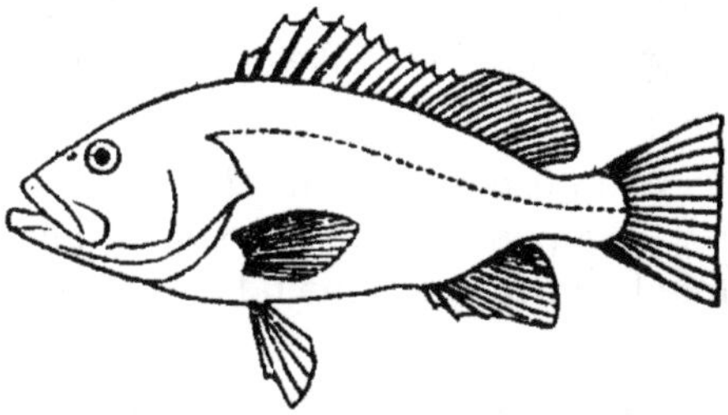

Figure 237: ***Epinephelus drummond-hayi***

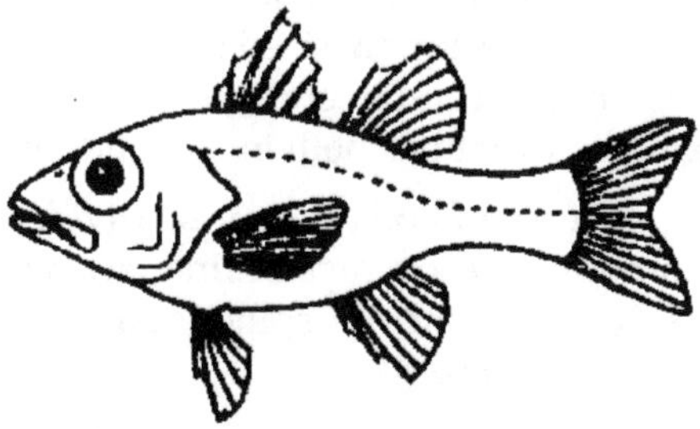

Figure 238: ***Apogon frenatus***

In a few fishes like weever (*Trachinus* species) and in poison fish the spine of the dorsal fin are associated with poison gland *(Figure 239).*

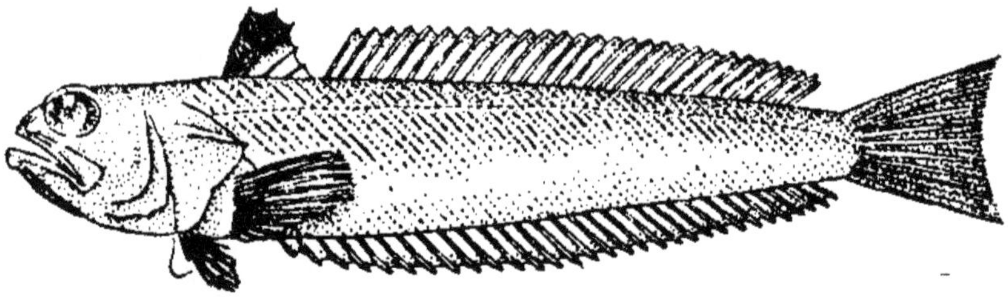

Figure 239: *Trachinus draco*

In majority of fishes, the dorsal and anal fins are capable of being erected or depressed at will. In the sail fish (*Istiophorus*) the spinous dorsal fin is of enormous size and is believed to be projected from the surface of the sea and used like a sail to assist progression, but the whole structure can be tucked away into a deep groove when not required *(Figure 240).*

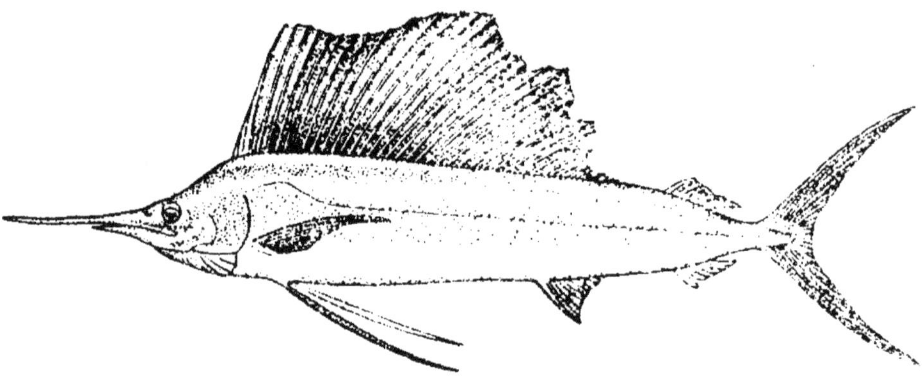

Figure 240: *Istiophorus americanus*

Anal Fin

It is present on the lower edge of the body behind vent. It is the least variable fin. It may be normal sized or greatly enlarged (*Balistes* sp., *Figure 241*) or elongated (*Figure 4*) or completely absent (*Gymnarchus* sp., *Figure 242*). Like dorsal fin it also exhibits variations both in size and form in different fishes.

In case of cods it may be divided into two portions (*Figure 215*). It may be supported by soft rays or first few rays may be converted into stiff spines (*Figure 40*). In some fishes the spinous part is separated off from the remainder as a distinct fin *(Figure 243).*

Anal fin may be modified into an intermittent organ called gonopodium for copulation in poecilinae (*Gambusia affinis*) (*Figure 244*).

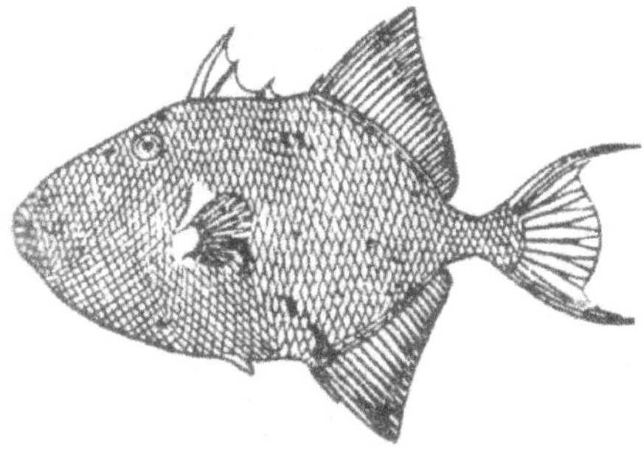

Figure 241: ***Balistes capriscus***

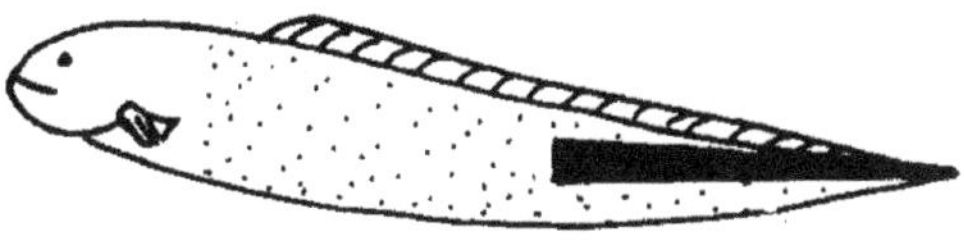

Figure 242: ***Gymnarchus niloticus* (anal fin is absent)**

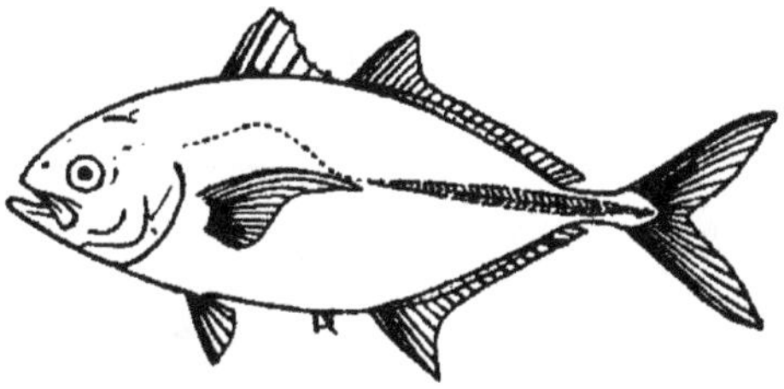

Figure 243: ***Caranx chrysos***

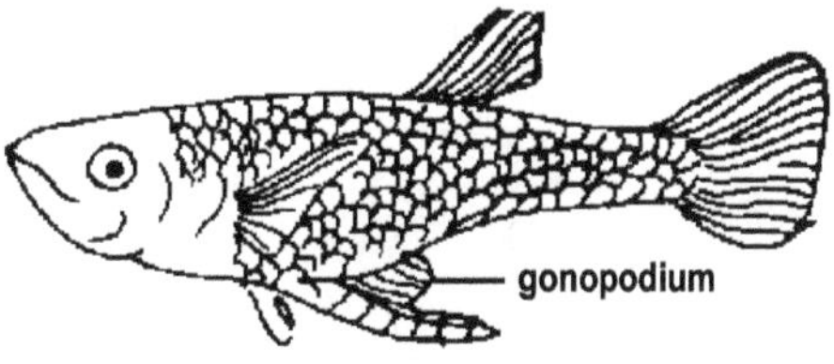

Figure 244: ***Gambusia affinis***

Caudal Fin

Like dorsal and anal, this fin is composed of simple or branched rays supported by a thin membrane. True spines are never developed in this fin but rudimentary or procurrent rays resembling spines may be found at the base of lobes in the nature of its supporting skeleton *(Figure 245)*. The shape of the caudal fin varies. It may be slightly emarginate or furcate, truncate, forked, notched, wedge or paddle-shaped, rounded, rounded with wavy margins, lanceolate or ovate [*Figure 246 (a–i)*].

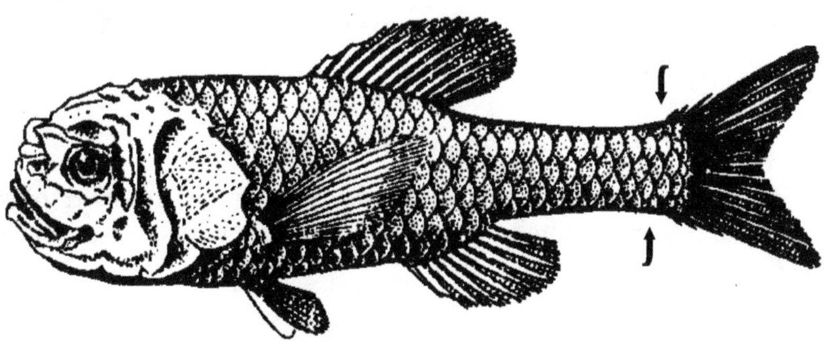

Figure 245: ***Melamphaes beanii***

Among the bottom dwelling rays (rajidae), the caudal fin tends to much reduced in size while in sting rays it is altogether absent, the whip like tail simply tapers to a fine point *(Figure 7a)*.

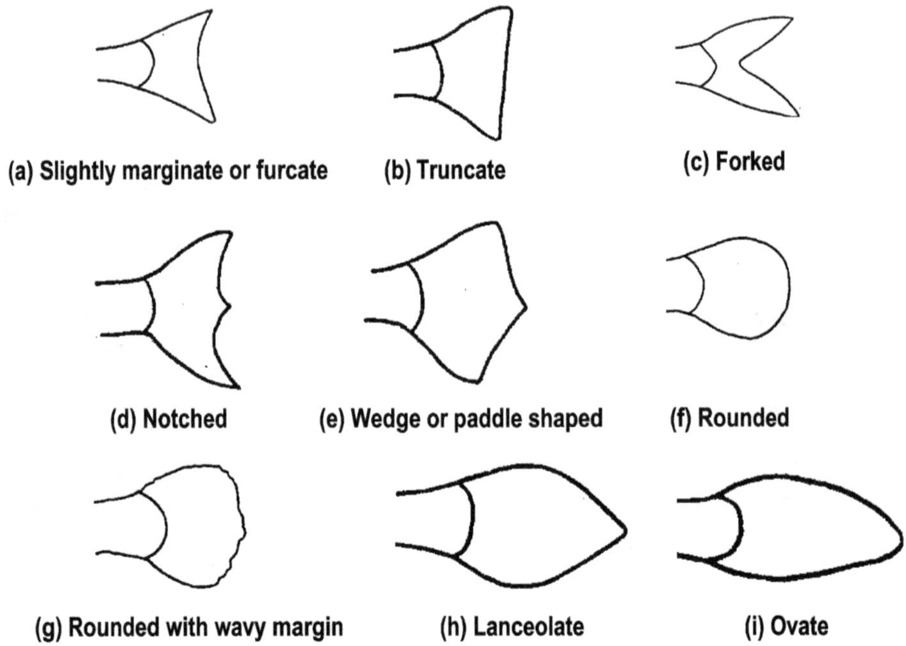

Figure 246(a–i): Variations in Shape of Caudal Fin

Some fishes may have unusual modifications. In *Trachypterus* sp., the caudal fin is directed upward at right angles to the axis of the body *(Figure 247)*.

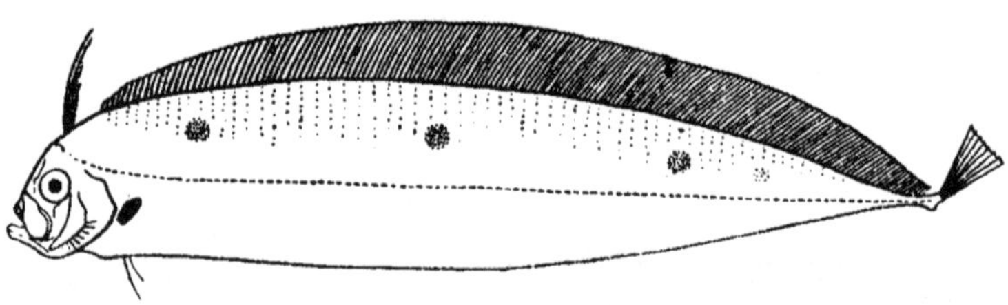

Figure 247: *Trachypterus arcticus* (Adult)

In sun fish (*Mola mola*), the body ends abruptly behind the short, high dorsal and anal fins and is marginated by low round, caudal with a slightly wavy edge (*Figure 248*).

Paired Fins

Pectoral fin: These are generally localised high up on the sides of fish and vary very little in position. Their shape varies to some extent. The shape of pectoral may be broad, rounded or spatulate *(Figure 249)* for the propulsion or backing up of water in the slow speed fishes. The fins may be longer and frequently sickle shaped (falcate) *(Figure 250)* in speedy fishes. Structurally it varies in different groups of fishes.

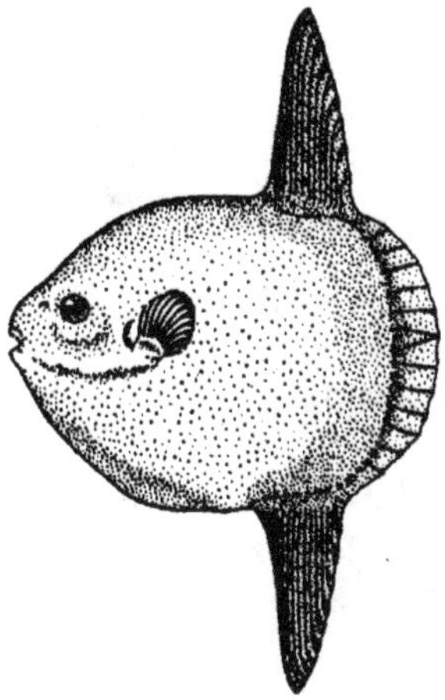

Figure 248: ***Mola mola***

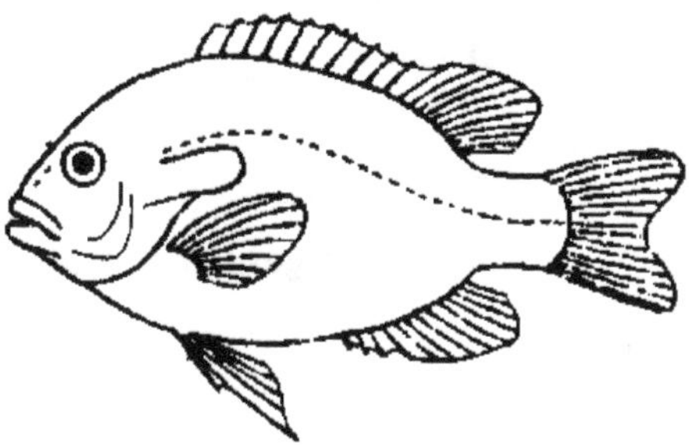

Figure 249: ***Lepomis megalotis***

The pectorals may become leaf-like in Australian lung fish (*Figure 251*) or other lung fishes from Africa and South America. The central lobe becomes long and narrow, the marginal fringe is reduced and fin takes the form of long, tapering filaments (*Figure 252*).

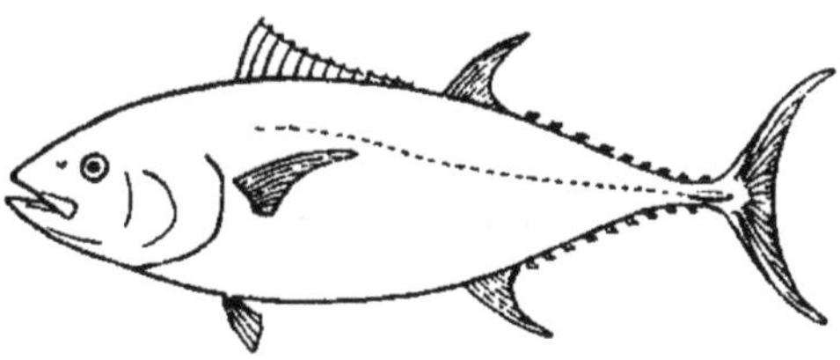

Figure 250: ***Thunnus thynnus***

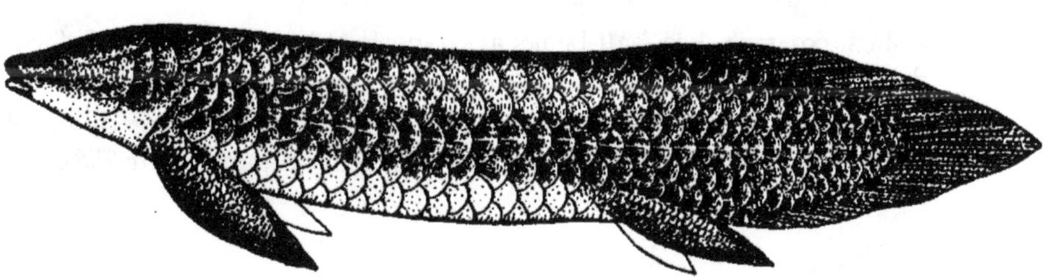

Figure 251: ***Epiceratodus forsteri***

In some pipe fishes and some eels, the pectoral fin becomes absent *(Figure 253)*. In sea toads and frog fishes (antenneriodae), these fins take the form of arms, ending in many fringed 'hands' by which they are able to crawl slowly on the sea floor *(Figure 254)*. In bat fishes (*Ogcocephalus* sp.), the arms are even more muscular *(Figure 255)*.

Pectoral fin undergoes adaptations in some cases:

- It may be enlarged in the form of a wing for flying *e.g., Exocoetus* sp. *(Figure 30)*.

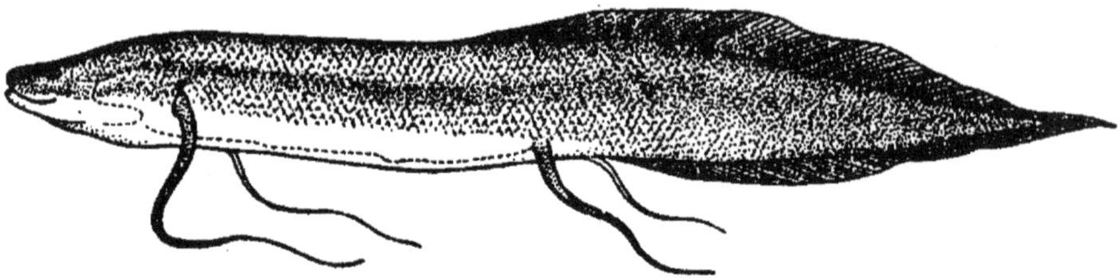

Figure 252: ***Protopterus aethiopicus***

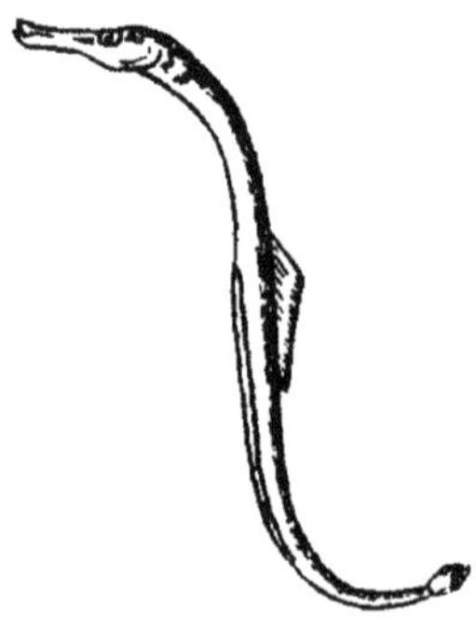

Figure 253: ***Siphostoma floridae***

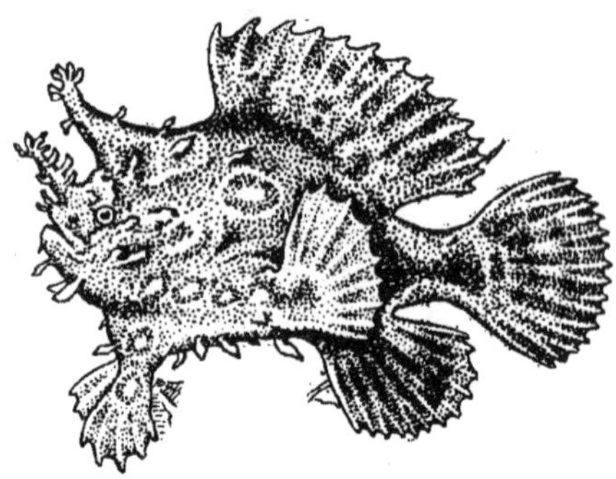

Figure 254: ***Pterophyryne tumida*** **(Frog Fish)**

- It may be muscular and feet like to enable the fish to hop *(Figure 256).*
- In some fishes, some or all rays of pectoral fin are drawn out into long and delicate filaments (*Polynemus* sp.) *(Figure 257).*
- Fishes like gar pikes, half-beaks are experts at leaps or "skittering" over the surface *(Figure 13).*
- In some fishes, the lower rays of the pectorals are prolonged, simple, thickened and free at their tips to act as sensory organ *(Figure 258).*

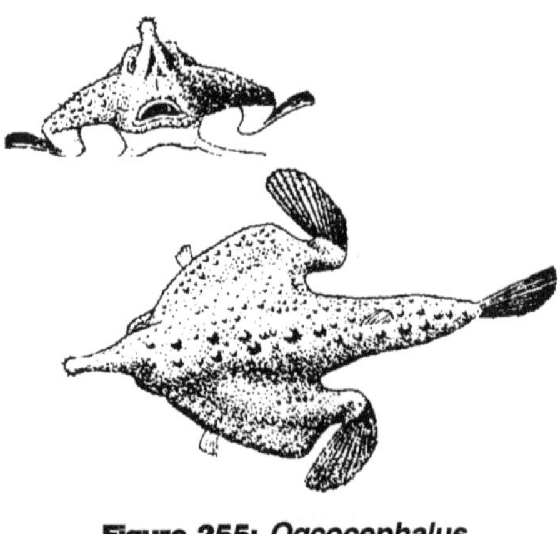

Figure 255: ***Ogcocephalus vespertilio*. (After Norman, 1984).**

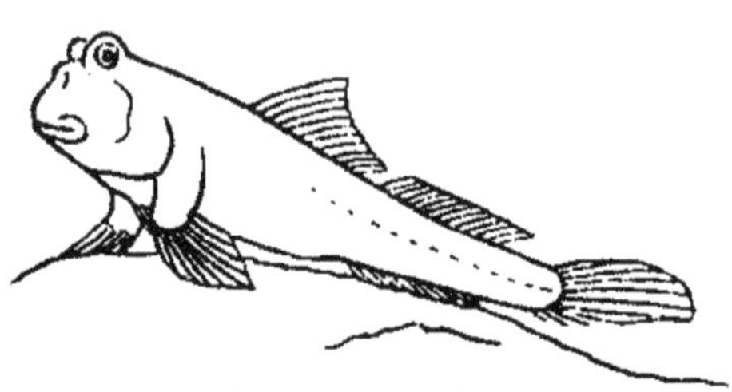

Figure 256: ***Periophthalmus koelreuteri***

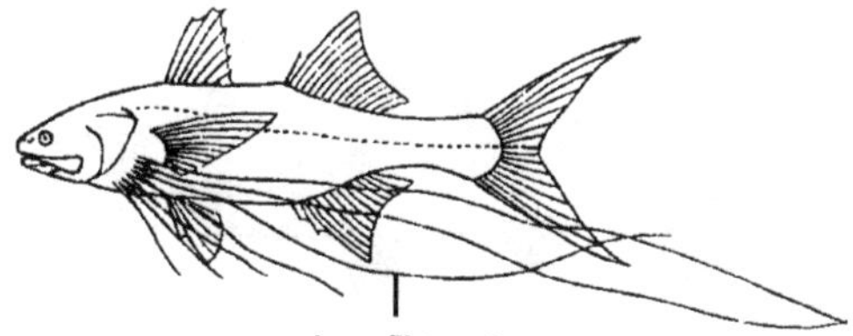

Figure 257: ***Polynemus paradiseus***

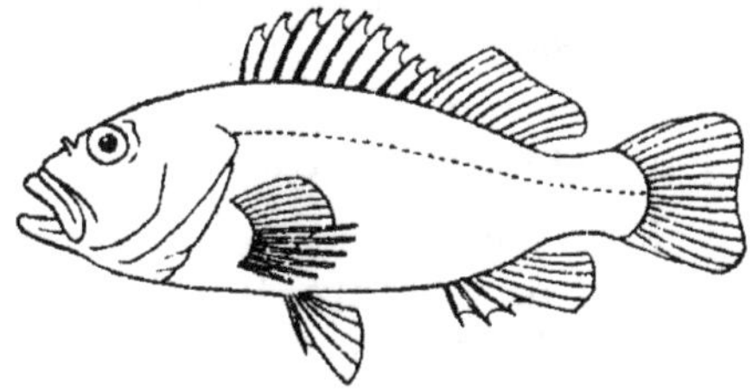

Figure 258: ***Paracirrhites forsteri***

- In some gurnards (*Trigla* sp.) and sea robins (*Prionotus* sp.), three of the lower rays of the pectoral fin are detached from the remainder and modified to form stout-finger-like appendages *(Figure 259)*.

Pelvic Fin

They are of considerable importance in the classification. The different positions of pelvic fins have been shown in [*Figure 260 (a–c)*]. In all the selachians, and in lower bony fishes such as herrings (clupea), salmon (*Salmo* sp.) and carps (*Cyprinus* sp.), the pelvics are placed in the middle of the belly between the pectorals and the anal and are said to be abdominal in position *(Figures 219, 261)*.

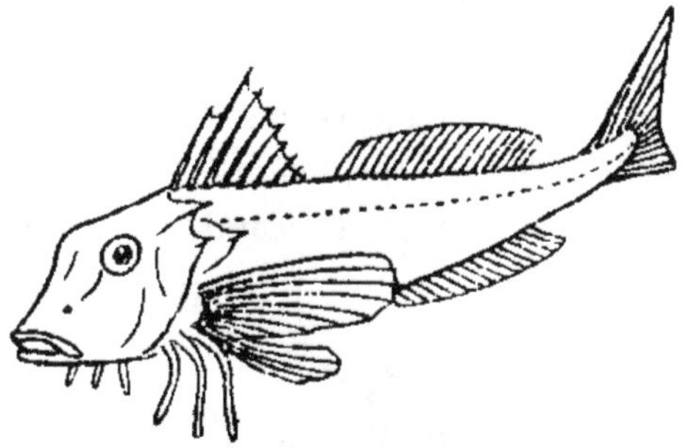

Figure 259: ***Trigla pini***

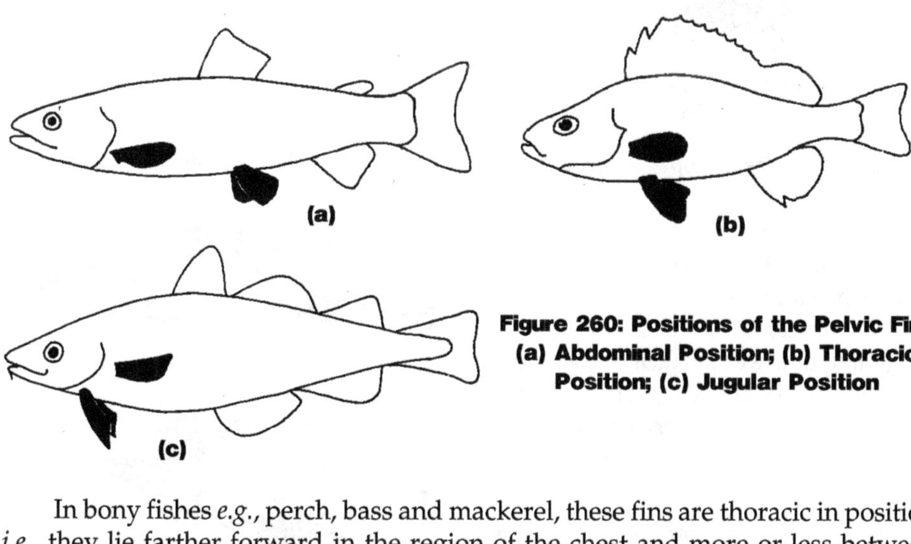

Figure 260: Positions of the Pelvic Fin– (a) Abdominal Position; (b) Thoracic Position; (c) Jugular Position

In bony fishes *e.g.*, perch, bass and mackerel, these fins are thoracic in position *i.e.*, they lie farther forward in the region of the chest and more or less between pectorals *(Figures 223, 249, 250)*.

In cods (gadidae) and certain blennies (blennioidea), they are jugular in position *i.e.*, they lie in front of the pectorals in the region of the throat *(Figure 215).*

In certain fishes (eels), they are much reduced or wanting. In pipe fishes, symbranchoid eels, gymnotids, globe fishes and porcupine fishes they are absent *(Figures 229, 242, 253).*

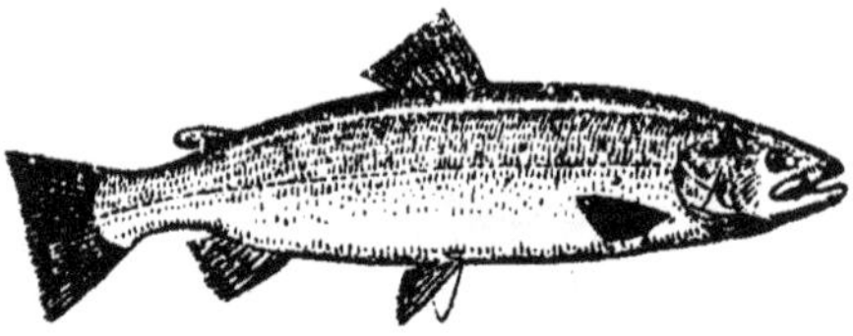

Figure 261: ***Salmo salar***

In cods (gadidae), these are reduced to filaments and in oar-fish (*Regalecus* sp.), pelvic fin is represented by a pair of long rays, each expanded into a blade like structure at the tip *(Figure 216).*

In sticklebacks (*Gasterosteus* sp.) each pelvic is comprised of a sharp-pointed spine and one soft ray *(Figure 118).*

It is reduced to a spine alone in pine-cone fish (*Monocentris* sp.) and trigger fish (*Plectograthi* sp.) *(Figures 240, 262).*

In File-fishes (monacanthidae), the pelvic bone with its spine is freely moveable *(Figure 263).*

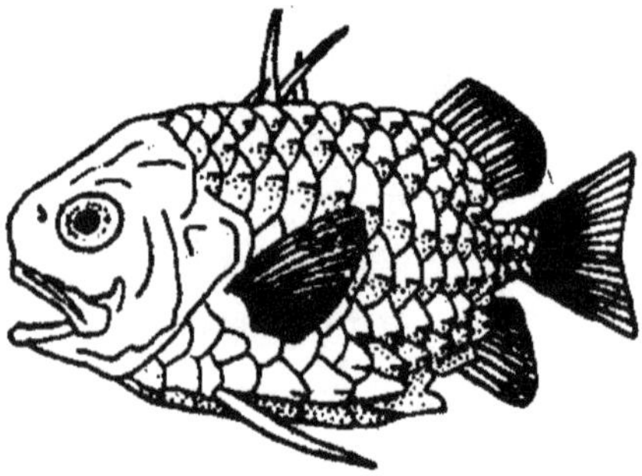

Figure 262: ***Monocentris japonicus***

Figure 263: ***Monacanthus*** **sp.**

The most important of the modifications is to form sucking disc to enable the fish to cling to rocks and stones. In little bornean sucker (*Gastromyzon* sp.), the whole of lower surface of the body is modified to form a large sucker *(Figure 264)*, where long and horizontally placed pectorals and pelvics play an important part.

In gobies (gobioidea), pelvics are united to form a deep cup-like sucker *(Figure 265).*

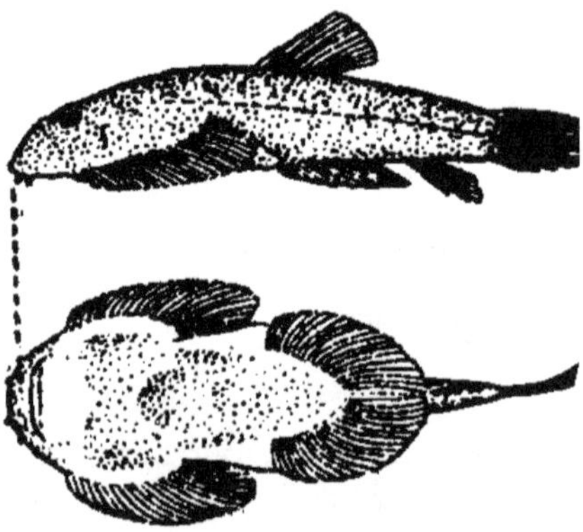

Figure 264: ***Gastromyzon bornunsis*** **(After Norman, 1984)**

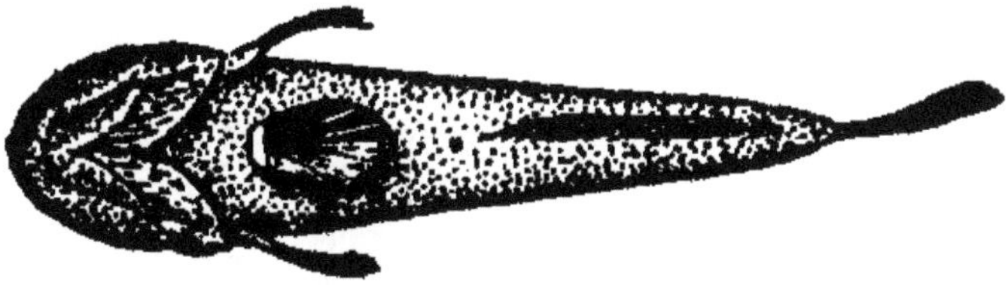

Figure 265: *Gobius niger*
(After Norman, 1984)

In lump-sucker (*Cyclopterus* sp.) and seasnail (*Liparis* sp.), a sucking disc is developed and pelvics lose their fin-like appearance (*Figure 266*).

In cling fishes (gobiesocidae), the disc is large and is composed of pads of thickened skin formed by the pelvic fins *(Figure 267).*

In sharks and rays, the hinder part of the pelvic fin in male are modified to form "claspers" that assist in the process of fertilisation *(Figure 268).*

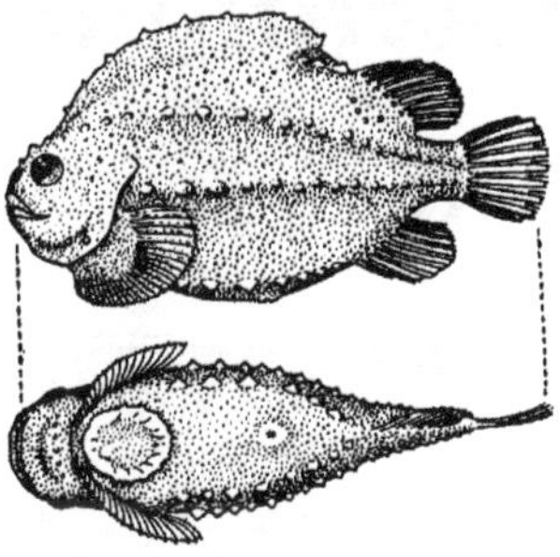

Figure 266: *Cyclopterus lumpus*
(After Norman, 1984)

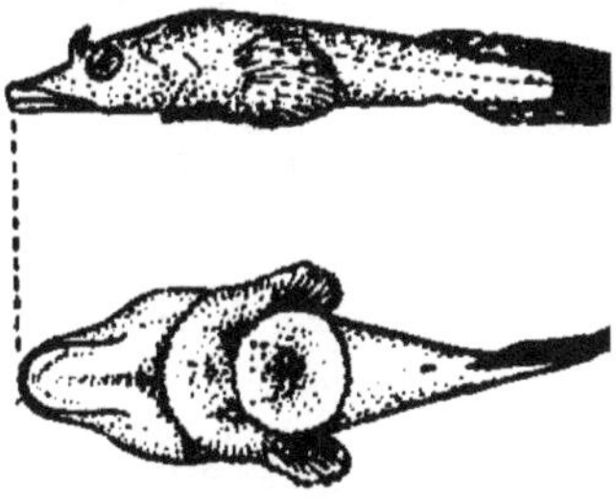

Figure 267: *Lepadogaster gouani*
(After Norman, 1984)

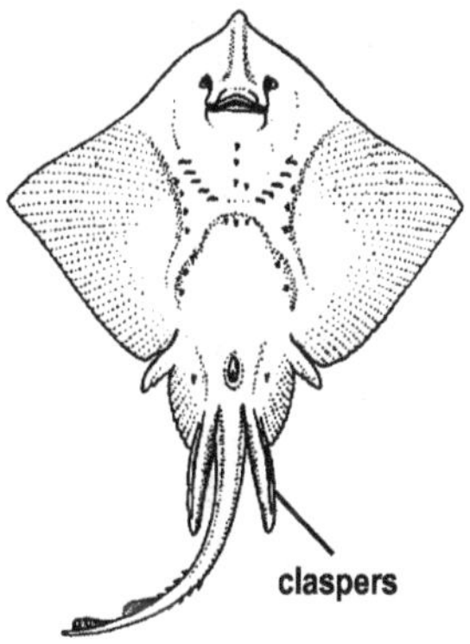

Figure 268: *Raia clavata*
(Male)

//••••••••••//

Chapter 9

Collection and Identification of Egg, Larvae and Juveniles of Some Common Fishes of India

Collection and transfer of spawn and fish fry from nature to culture farms is an age-old practice and is in vogue even today when the fish biologists have learnt the technique of induced breeding. The induced breeding though has revolutionised the fish culture practices but its applicability is still limited to very few freshwater fish being cultured in India or abroad. The majority of fish farmers in rural areas still believe in collecting spawn from nature or at the best purchasing it from the fish traders who are mainly concerned with seed production. In both these situations a culturist depends for seed either on nature or the seed supplier and often suffer because the seed supplied is of mixed quality and thus a good part of his culture efforts go in vain as at the end he finds a substantial volume of produce being of raw forage fish, least accepted in the market. Thus it is of extreme importance that spawn collected from nature or even purchased could be tested for its purity or some mechanism developed for segregation through proper identification and selection.

Collection

Sources of Seed

Fish seed can be collected from nature, confined and running waters as per the breeding behaviour of fish.

In confined waters, like ponds/lakes, it is very easy to collect seed, whereas in flowing waters, the collection of spawn is very difficult. The major carps of India–rohu, catla, mrigala, tor etc. breed in rivers during monsoon and collection of their spawn is very difficult and laborious process. Moreover, it is also mixed with the spawn of undesirable ones.

Seed Collection from Rivers

Once the breeding grounds are identified, eggs are collected by disturbing the bottom and scooping them with a "gamcha". In Bihar, two breeding grounds have been located, one at river Badua and other at Dingraghat on Mahanadi; in Gwalior at Banmore; and in Jhoroni (Morene) at Madhya Pradesh.

On river Ganga, Yamuna, Betwa, Gomti, Ramganga, Ghaghra etc., these collection centres are located at Agra, Etawah, Fatehpur, Allahabad, Lucknow, Moradabad (Taipur), Salempur, Bareilly (Uttar Pradesh).

Fish seeds are collected from shallow margins, especially where there is a curve in the course of river.

The spawn collection net is popularly called as "Benchi jal" or shooting net. It is funnel-shaped, made of closed mesh or cheap coarse cloth. Its length varies from 15–22 feet (*Figure 269*). Its both ends are open. The mouth is much larger than the tail. The net is fixed to bamboo poles in shallow margins of flooded river, with the mouth facing the current. The net is allowed to drift in the water so that it positions itself along the direction of the current. The net is then stretched along its axis and posterior end is also tied to a bamboo pole. At the posterior end of the net there is usually a stitched in a ring of split bamboo or cane and to this is attached during operation, a tail piece called "gamcha".

The "gamcha" is often a rectangular piece of cloth of 2 × 1 × 1 feet. The anterior end of gamcha is tied round the ring of the net while the posterior end is fixed. Several such nets are operated at a time. Eggs, fry and hatchlings move along with the current and get collected in the "gamcha" from where they are periodically scooped and transferred either to happas or hatcheries. The fry should be removed from the "gamcha" carefully.

Fish seed of desirable species of carp can be identified by following features:

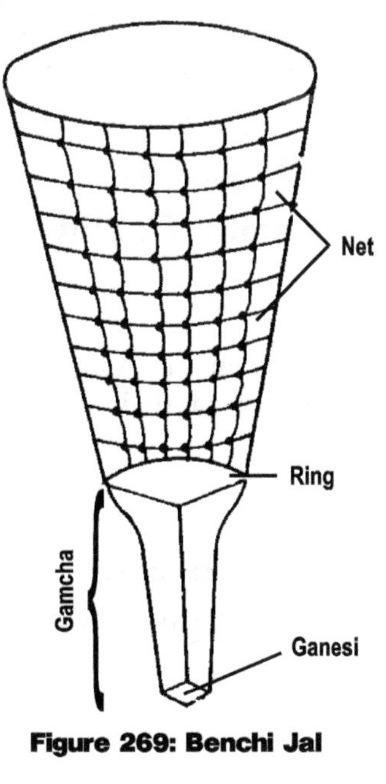

Figure 269: Benchi Jal

1. Colour of body, shape and size of eye and fins, presence of pigment spots, shape and size of air-bladder in advanced fry.
2. Shape and size of yolk sac, location of anal opening, eye pigmentation, opening of mouth, nature of the fin folds are helpful characters for identifying young fry of different species.
3. Another method is based on difference in the demand of DO by fry of difference species. When mixed collection of fry are forced to overcrowding, they segregate to different layers of water (surface, middle and bottom) according to their oxygen requirement.
4. Fertilized ova of different species of fish usually differ in their diameter. Sieving of spawn, using sieves of different grades, is used to separate the required species from the rest.

Preservation

For preservation of eggs, they are put in 10 per cent formalin containing glacial acetic acid (2.5 per cent) or hydrogen peroxide (2.5 per cent). For better understanding of embryonic stage in the egg, the eggs are treated with a mixture of glacial acetic acid (3.5 ml) and 0.7 per cent NaCl solution (100 ml). This makes shell transparent, embryo appears white and yolk becomes coloured.

Condition

The condition of egg is also important in the study. The healthy and fertilized egg attains the average diameter of the egg for the species, looks transparent, turgid and hard.

The dead or unfertilized eggs appear opaque, smaller in diameter and irregular in shape. The eggs collected from nature are of three types *viz.*

1. *Planktonic egg*: These eggs float passively.
2. *Semibuoyont egg*: These eggs neither float nor settle on bottom.
3. *Demersal egg*: These eggs sink to bottom.

Commonly Employed Terminology in Developmental Stages of Fish

These uncertain terms which are often used to specifically indicate the stage of development from egg to adult (developmental stages) and need to be well understood for their appropriate use:

Larva

A term used for any stage from hatching to metamorphosis but bears yolk sac.

Hatchling or Sac Fry

A just hatched stage with large amount of yolk but mouth still not formed.

Fry

A stage with open mouth for feeding and yolk present though diminishing fast.

Table 5: Some Identification Characters of Eggs of Marketable/Culturable Fishes of India

Sl.No.	Name	Colour	Shape	Diameter (mm)	Nature of Egg	Place of Breeding
1.	*Clarias batrachus*	Green	Oval	1.5 × 1.3	Adhesive	–
2.	*Heteropneustes fossilis*	Greenish	Round	1.5	-do-	Breeds in confined water.
3.	*Wallago attu*	Yellowish	-do-	1.2–1.5	-do-	–
4.	*Notopterus notopterus*	-do-	-do-	3.5	-do-	–
5.	*N. chitala*	-do-	-do-	6.5	-do-	–
6.	*Mastacembelus* spp.	Light brown	-do-	2.0	-do-	–
7.	*Channa punctatus*	Golden brown	-do-	1.0	Floating	Breeds in confined water.
8.	*C. marulius*	Amber	Oval	1.5–1.7	-do-	-do-
9.	*C. striatus*	-do-	Round	1.25–1.35	-do-	-do-
10.	*Anabas testudineus*	Transparent	-do-	0.7	-do-	-do-
11.	*Catla catla*	Yolk light red	Spherical	2–2.2 and increases to 4.4–5.2 when swollen-up	Non-floating	Deposits eggs in shallow marginal areas of rivers
12.	*Labeo rohita*	Yolk reddish	Round	5.0	Non-floating	Shallow indented marginal ares of rivers
13.	*L. calbasu*	Bluish	Oval	5.3–6.0	-do-	Flooded river margins
14.	*L. gonius*	Bluish	Round	4.5	–	–
15.	*L. fimbriatus*	Bluish tinge	-do-	3.5	–	–
16.	*L. kontius*	Faint pale blue	-do-	3.8	–	–
17.	*L. dero*	Colourless	–	2.9–3.2	–	–
18.	*Hilsa hilsa*	Segmented with numerous oil globules	-do-	2.1–2.3	–	Breeds in flooded waters

Contd...

Table 5–Contd...

Sl.No.	Name	Colour	Shape	Diameter (mm)	Nature of Egg	Place of Breeding
19.	*Ctenopharyngodon idella*	–	–	–	–	Eggs are deposited in middle of the river in rapid current
20.	*Cyprinus carpio*	–	–	1.9	–	Breeds in confined water
21.	*Schizothorax plagiostomus*	Orange	–	3.0–3.3	–	Breeds in flowing water
22.	*S. planifrons*	Lemon	–	3.0–3.1	–	–
23.	*S. curvifrons*	Yellow orange	–	3.2–3.4	–	–
24.	*S. longipinnis*	Orange translucent in appearance	–	3.6–3.8	–	–
25.	*Tor putitora*	Bright yellow	–	3.2–3.1	–	–
26.	*Barilius bendelosis*	Pink to pale yellow	–	1.6–2.0	–	–
27.	*Barilius shacra*	Lemon	–	2.2–2.6	–	–
28.	*Glyptothorax kashmirensis*	Crimson red	–	2.2–2.3	–	–
29.	*G. pectinopterus*	Green	–	2.1–2.3	–	–
30.	*Nemacheilus corica*	Off-white	–	2.4–2.8	–	–
31.	*Cirrhinus mrigala*	Brown yolk	Round	1.5 and upon swelling measures 3-4 mm	Non-floating	Breeds in rivers

Post Larva

A stage with no yolk, but structurally unlike the young.

Alevin or Advanced Fry

A stage from complete absorption of yolk to a size of 1 inch in length, and resembling the young in structure.

Fingerling

A stage beyond advanced fry to a size of 5 inches in length.

Young or Young-of-the-year

A stage from hatching to January 1 in Northern Hemisphere or July 1 to the Southern Hemisphere, representing age group 0.

Yearling

A stage at age group 1 *i.e.,* in the second year of life.

Juvenile

A stage anywhere between young and the adult.

Year Class

A group of fish that spawned and hatched in same calendar year.

Recruit

A stage of younger fish when it becomes vulnerable to fishery and contributes to adult population size.

Consolidated data (following Jhingran and Sehgal, 1978; Jhingran, 1991 and Srivastava, 2000) on egg, larvae and their juveniles has been depicted in Tables 5, 6, and 7 and Figure 270.

Table 6: Postlarval Stages of Fish Families

Cyprinidae	Eyes pigmented, air bladder conspicuously dark, embryonic fin-fold continuous, anal opening in the posterior 1/3 of the body, mouth formed, pectoral fins enlarged, barbels absent or present.
Siluridae	Eyes pigmented and small, 1-4 pairs of well developed barbels, air bladder absent, embryonic fin-fold continuous, anal opening near about the middle of the body, mouth formed.
Clupidae	Body slender and compressed, head obtusely pointed, mouth formed and terminal, pectoral fins differentiated, eyes large and pigmented, dorsal fin differentiated, larval fin-fold discontinuous, anal opening in the posterior 1/3 of the body.
Channidae	Eyes pigmented, pectoral fins well developed, air bladder present, embryonic fin-fold continuous, yellow and dark horizontal bands along the sides, barbels absent.
Notopteridae	Maxillary barbels present, pelvic fins joined to a certain extent along the inner side from the base and free distally, a number of vertical bands.

Contd...

Table 6–Contd...

Gobiidae	Eyes pigmented, barbels absent, pectorals functional, anal opening near about the middle of the body, air bladder conspicuous, embryonic fin-fold continuous, a large pigment spot above the air bladder.
Mugilidae	Barbels absent, first dorsal fin indicated as a narrow strip of the embryonic fin-fold, second dorsal well differentiated, anal also differentiated, ventral fin rudimentary, anal opening near about the middle of the body.
Sciaenidae	Eyes pigmented, barbels absent, anterior portion of boy deeper than caudal, indication of rays in the fin-fold very distinct on the dorsal side, around the caudal and on the ventral side, pre-anal fin-fold already disappeared, indications of anal fin formation appear as a thickening of the continuous median fin-fold, anal opening centrally situated.
Cichlidae	Eyes large and pigmented, barbels absent, anal situated in the anterior half of the body, embryonic fin-fold continuous and narrow.

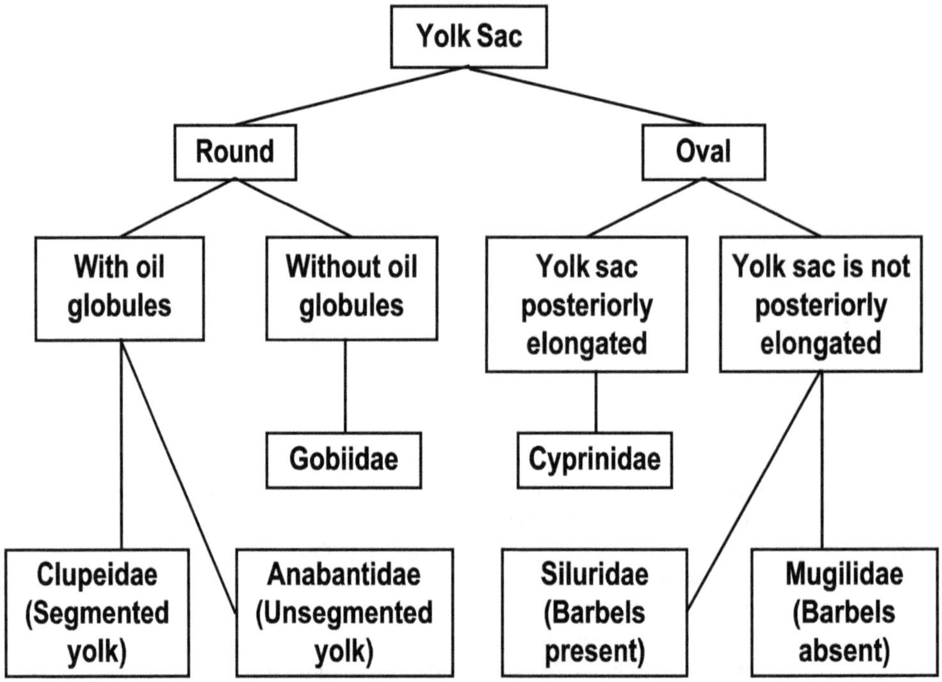

Figure 270: Identification of Prolarval Stages of Fish Families

Table 7: Distinguishing Features of Larvae, Post-larvae and Juveniles of Certain Coldwater Fishes

Species	*Larva*	*Post-larva*	*Juvenile*
Tor putitora	Average total length 13.0 mm, scattered pigment on the back extending upto caudal fins, conspicuous black spot at the base of tail fin, two pairs of barbels.	Average total length 9.7 mm, mouth clefted extending upto the front portion of eyes, dorsal fin with 2 strong and 9 soft rays; base of tails has a black spot.	Average total length 5.4 mm, bright yellow red pigment on the yolk-sac, eyes non-functional, 26 body segments.
Labeo dero	Average total length 14.7 mm, pigment on the head extend upto the origin of dorsal fin, thick and continuous lips present, two pairs of barbels.	Average total length 9.9 mm, dorsal fin with 8 rays, dark black blotch at the base of tail.	Average total length 4.1 mm, eyes non-functional, 25 body segments.
Barilius bendelisis chedra	Average total length 14.6, body elongate, caudal fin rounded. Clusters of pigment on back and parallel rows on mid-lateral side.	Average total length 5.2, ventral, dorsal and anal fins distinct, six large post-anal chromatophores.	Average total length 4.4 mm, body of slender, 18 body segments.
B. shacra	—	Average total length 5.3 mm, dorsal and anal fins rudimentary, 5 groups of pigment on dorsal side.	Average total length 4.09 mm, transparent and slender, eyes well-developed, 15 body segments.
Schizothorax plagiostomus	—	Average total length 10.7 mm, dorsal fin with 8 soft rays, black spot present on lower part of caudal peduncle, pigment scattered on dorsal lateral and ventral aspects of the body.	Average total length 5.4 mm, yolk-sac oval, 18 body segments.
S. planifrons	—	—	Average total length 6.6 mm, yolk-sac elongated, eyes functional, 22 body segments.
S. curvifrons	—	Average total length 9.4 mm, dorsal fin without rays, rows of pre-and-post-anal pigment present.	Average total length 7.2 mm, mouth placed on the lower side, 28 body segments.

Contd...

Table 7–Contd...

Species	*Larva*	*Post-larva*	*Juvenile*
S. longipinnis	Average total length 8.1 mm, yolk-sac pear-shaped, eyes functional, 29 body segments.	Average total length 10.5 mm, all fin lobes without rays, 4-6 large post-anal clusters of pigment, rudiment pair of maxillary barbels present.	—
Glyptothorax kashmirensis	Average total length 4.8 mm, yolk-sac pear-shaped, mouth ventral, 25 body segments.	Average total length 7.3 mm, mandibular and maxillary pairs of barbels present, eyes dorsal, adhesive disc on ventral side reduced, pigment scattered all over the body giving dull gray colour.	—
G. pectinopterus	Average total length 4.2 mm, yolk-sac pear-shaped, mandibular and maxillary pair of barbels present, body black, 19 body segments.	Average total length 7.3 mm, second pair of mandibular and one pair of rostral barbels present, rudiment rhomboid-shaped disc present.	Average total length 10.5 mm, teeth on upper and lower jaws, all fins flat, tail sharply bifurcated.
Nemacheilus corica	Average total length 3.6 mm, yolk-sac pear-shaped, eyes non-functional, 22 body segments.	Average total length 8.9 mm, mouth ventral, maxillary and mandibular pairs of barbels present, clusters of pigment at mid-lateral parts of the body.	Average total length 14.6 mm, mouth ventral and outer margin fringed, back with irregularly arranged bands and spots.

Chapter 10

Freshwater Food Fishes of India and Determination of their Freshness

Fish has attracted man since times immemorial for its nutritive, medicinal and recreational values. Fresh fish flesh provides an excellent source of protein for human diet and is superior to milk, beef protein and egg albumen. Fish flesh, therefore, becomes a valuable supplement to human diet for people who are habitually taking cereals, starchy roots and sugar as their principal diet. About 300 g of raw fish flesh suffices to supply about half of the total protein and fat requirement and one-fourth of the calories needed in a balanced daily diet for adult human beings. Also, this amount of raw fish provides 50 per cent of phosphorus, 30 per cent of iron, 100 per cent of vitamin A (fat fish only), 30 per cent of vitamin B_1, vitamin B_2 requirement and 50 to 100 per cent of niacin requirement. The digestibility of fish flesh is in the order of 96 per cent.

Fish is an effective supplementary food where diet is principally plant oriented. Today, preference of fish as food is not because of any legacy of the past but on its own merits, as judged by present standards of nutrition requirements and dietary allowances. Fish is a good substitute for the pulses and so can lessen the pressure on agriculture to some extent. Eating fish once a week is also recommended by FAO.

The fish industry in some states of India *viz.*, West Bengal, Bihar and Orissa is about 1500 years old. In West Bengal, every family traditionally has atleast one small or moderate size fish pond. The increased demand of fish concomitant with increase in population is exerting an unabating pressure on capture fishing thus threatening the existence of most sought for varieties of edible fish. Needless to say, a good supply of fish through sustainable capture/culture fishery can go a long way to overcome the human food problem that has emerged in recent years as a result of exponential growth in urbans.

Thus, the increasing market demand for fish and for consumer awareness there is a need to educate both marketing folk and the consumer about the means to know local name and identify fish, to ascertain preferred edible varieties and also to know sufficient ways to understand freshness of fish being marketed or purchased. In India, majority of food fishes are teleosts, except some sharks that belong to the order/families of superorder teleostei of sub-class teleostomi.

Key to Identification of Indian Edible Freshwater Fishes

When Abdomen is Keeled and Serrated (Order: Clupeiformes), they may be Estuarine or Freshwater Inhabitant *e.g.*

Hilsa ilisha

Popularly known as hilsa (Bengali–ilisha; Telegu–palasah; Marathi–pala; Tamil–Oolum; Kannada–paliya; Malayalam–paluva; Mumbai–chaksi; Uttar Pradesh and Bihar–hilsa), it is very popular, edible and delicious fish that is rich in the oils. Mouth is large and pelvics are very small *(Figure 271)*. The fish is caught in Bengal, Orissa and Kathiawar coasts. It migrates into rivers–Ganges, Mahanadi, Hoogly etc., and is caught and marketed almost year around.

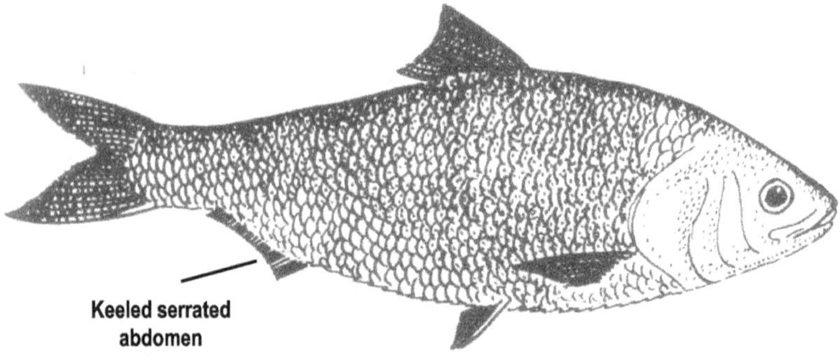

Figure 271: ***Hilsa ilisha***

Genus–*Salmo*

It is characterised by elongated subcylinderical smooth and rounded abdomen, obtuse snout, large eyes and adipose fin. Introduced in cold waters of J&K and Karnataka states and is a good sport fish rarely marketed. Various species of *Salmo* are:

Salmo gairdneri gairdneri (Rainbow Trout)

Body is without spots below the lateral line, colour steel blue and the lateral line scales are 127-160 (*Figure 20*).

***Salmo trutta fario* (Brown Trout)**

Body is characterised by many coloured spots above and below lateral line, adipose fin having an orange tinge. Body is coloured brown. Lateral line scales are 115-130. In Himalayan region, it is the only trout that supports trout-fishing while in the Deccan Plateau, the irideus and shasta strains are the principle ones (*Figure 272*).

Figure 272: ***Salmo trutta fario***

***Salvelinus fontinalis* (Eastern Brook Trout)**

This species once very popular is fast losing ground because of its poor adaptability. It has the fastest growth rate among the three species and has the best flesh. This fish has poor tolerance to temperature above 18^{0}C and less resistant to diseases and pollutants. Its body is devoid of spots but mottled or barred, with dark olive colour with a reddish band along the sides in male; dorsal fin with black marks on a paler background. Anterior margin of the anal fin is snow-white with thin streaks of white on other fins (*Figure 273*).

Figure 273: ***Salvelinus fontinalis***

Genus–*Notopterus*

Body is oblong, compressed and deep. Caudal region is long and tapering. Abdomen with a double series of 28-46 scutes. Anal fin is very long with 100-135 rays and is confluent with caudal fin. Adipose fin is absent. Important species are:

Notopterus notopterus

Its local name is 'kapti'. In this opercular scales are much larger than those on body. No black spots on the tail end. Dorsal part of the body less humped. 28-33 scutes along the abdominal edge. Head is small and dorsal fin sticks out like a feather and hence popularly named as "feather backs" *(Figure 274).*

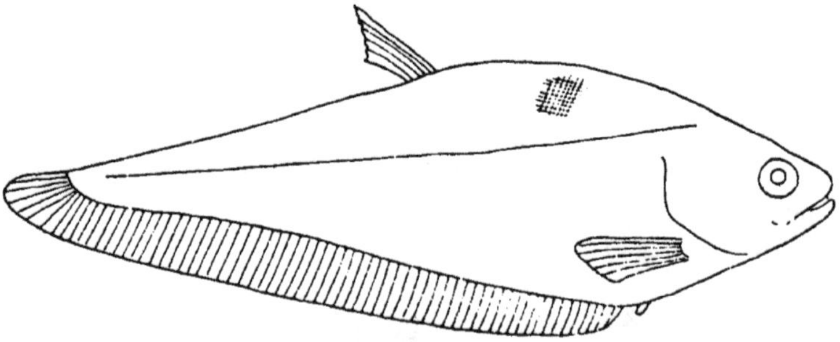

Figure 274: ***Notopterus notopterus***

Notopterus chitala

Its local name is moya. Bengal–chital; Oriya–pulli; Bihar–mohi; Punjabi–moh; Uttar Pradesh–moya. The opercular scales are larger than those on the body. Maxilla extending for beyond the hind edge of the orbit. A few black spots at the tail end. Dorsal part of the body is much humped. 50-51 scutes along abdominal edge (*Figure 275*).

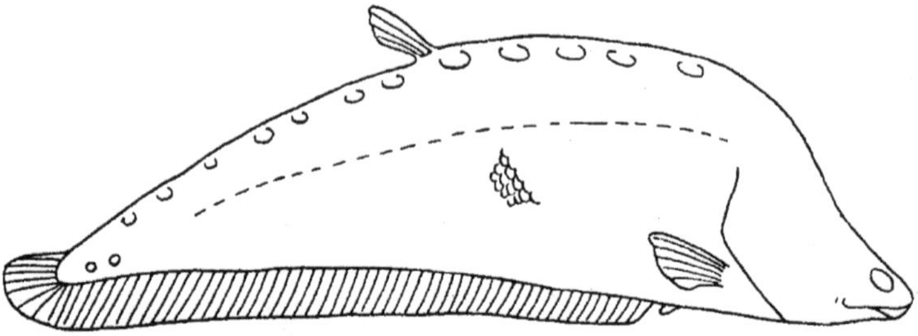

Figure 275: ***Notopterus chitala***

When Abdomen is Keeled or Rounded but not Serrated (Order: Cypriniformes)

It includes all major and minor carps, cobitids and cat fishes (Table 8).

Genus–*Labeo*

Some of the species of *Labeo* are:

L. rohita

Its local name is rohu. Bengali–ruee; Marathi–tambada masa; Oriya–rohu; Punjabi–rohu; Assamese–rohitii. It is present in the freshwaters of Northern India

Table 8: Some Characteristic Features to Identify Different Genera of Fishes

Ctenopharyngodon	*Labeo*	*Cirrhinus*	*Catla*	*Cyprinus*	*Tor*	*Hypophthalmichthys*	*Schizothorax*	*Schizothoraichthys*
(*i*) **Body** is moderately elongated, subcylindrical anteriorly and compressed posteriorly. Operculum is large and square. (*ii*) **Head** is depressed and flat. (*iii*) **Lips** are thin. (*iv*) **Dorsal fin** is inserted slightly ahead of pelvic fins nearer tip of snout than caudal fin base with 10 rays (3 simple and 7 branched) and without a spine. (*v*) **Scales** are large and cycloid.	(*i*) **Body** is small or moderately sized, elongated to deep with abdomen rounded. (*ii*) **Lips** are thick, flesh, fringed and form a labial fold. (*iii*) **Dorsal fin** has 3 spines and 13 rays. Caudal fin is deeply forked or emarginate. Dorsal fin is inserted above anterior to origin of pelvic fins with 11-26 rays and without osseous ray.	(*i*) **Body** is moderate, elongated and compressed. (*ii*) **Upper lip** and **lower lips** are not continuous. (*iii*) **Fins** are orange with black margins. (*iv*) **Lower jaw** is sharp with small tubercles at the symphysis. (*v*) **Lateral line** with 35-45 scales. (*v*) **Dorsal fin** is inserted ahead of pelvics with 10-19 rays and without spines.	(*i*) **Body** is short and deep. (*ii*) **Snout** is blunt. (*iii*) **Head** and **eyes** are large. (*iv*) **Upper lip** is absent and **lower lip** is thick. (*v*) **Dorsal fin** is long, inserted above tip of pectoral fin with 18-19 rays and without any spine.	(*i*) **Body** is robust anteriorly, more or less compressed. (*ii*) Pairs of barbels. (*iii*) **Lateral line** scales 30-40. (*iv*) **Dorsal fin** is very long and inserted above tip of pectoral fin with 3 spines and 17 rays. Third spine is strongly serrated.	(*i*) **Body** is elongated and moderately compressed. (*ii*) **Head** is small and broadly pointed. (*iii*) **Scales** are large. (*iv*) **Snout** is angularly rounded with tubercle. (*v*) **Dorsal fin** is inserted above pelvic fins with 12-13 rays and a strong stout smooth spine. (*vi*) **Lateral line** scales 22-30.	(*i*) **Body** is stout and compressed. (*ii*) **Abdomen** is strongly compressed with sharp keel from breast to vent. (*iii*) **Snout** is blunt and round. (*iv*) **Lips** are thin and lower jaw with tubercle.	(*i*) **Mouth** is more or less crescentic. (*ii*) **Lower lip** is modified into a papillated sucker. (*iii*) **Scales** are small vent and base of anal fin enclosed by a tile like row of scales.	(*i*) **Body** is elongated, subcylindrical with both profiles arched. (*ii*) **Snout** is smooth, obtuse and in some studded with sharp wart like protuberances. (*iii*) **Barbels** are 4 in number. (*iv*) **Lower lip** is non-suctorial. (*v*) **Dorsal fin** is inserted opposite pelvics with 11-12 rays and one spine.

and upto Godavari river in Central India. It has also been introduced in the Peninsular India. It is one of the finest of food fishes *(Figure 276)*.

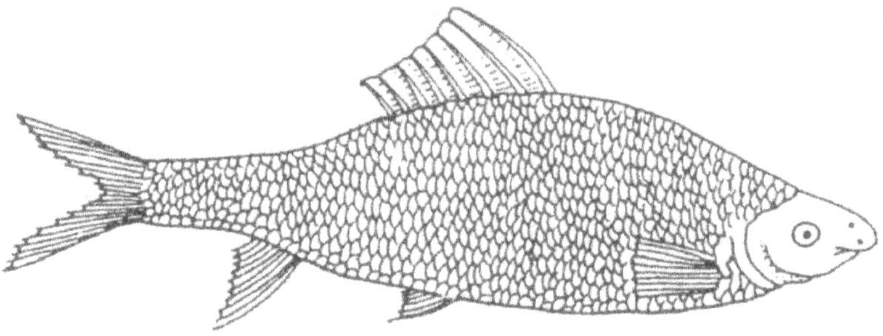

Figure 276: ***Labeo rohita***

L. angra

Its local name is thuthunahia raia. It is distributed in Uttar Pradesh, Bihar and Orissa *(Figure 277)*.

L. bata

Local name is bata. It is found throughout India except Kerala *(Figure 278)*.

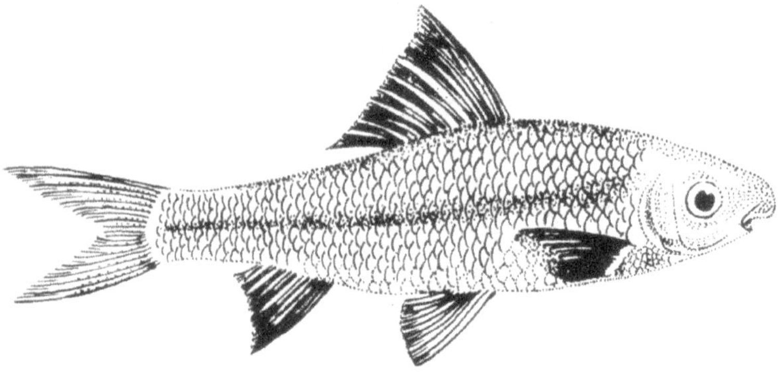

Figure 277: ***Labeo angra***

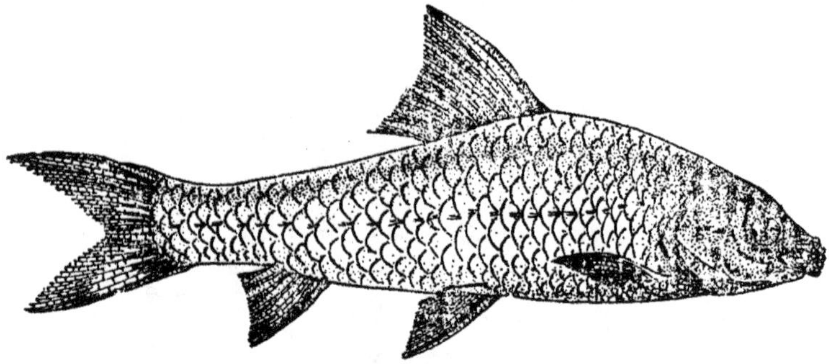

Figure 278: ***Labeo bata***

L. boga

Local name is bhagan. Except Kerala, it is present throughout India *(Figure 279).*

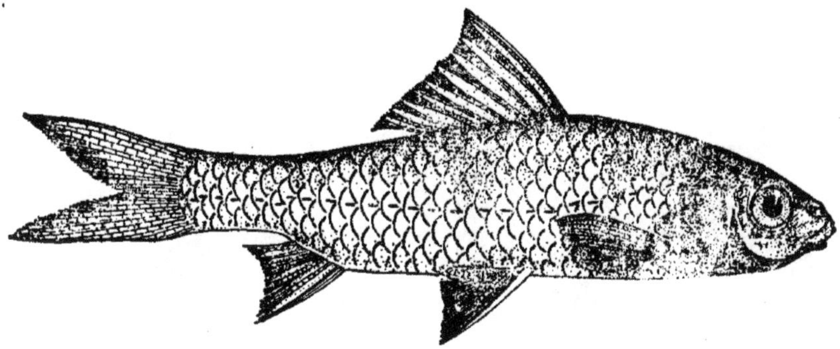

Figure 279: ***Labeo boga***

L. calbasu

Local name is karaunchar. It is distributed almost throughout India *(Figure 280).*

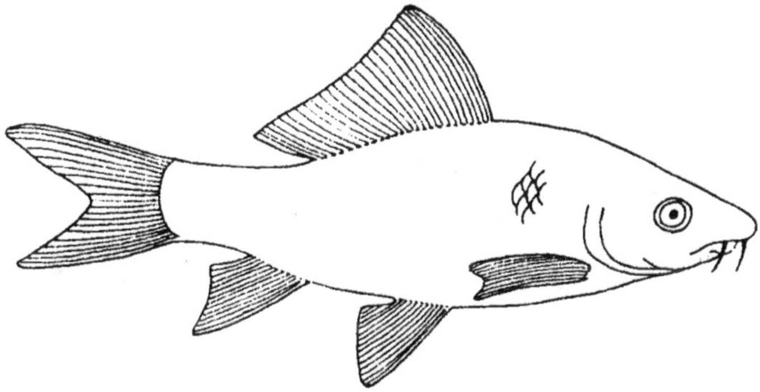

Figure 280: ***Labeo calbasu***

L. dero

Local name is gola raia, arangi, basanahawa raia. It is present throughout India, except Kerala *(Figure 281).*

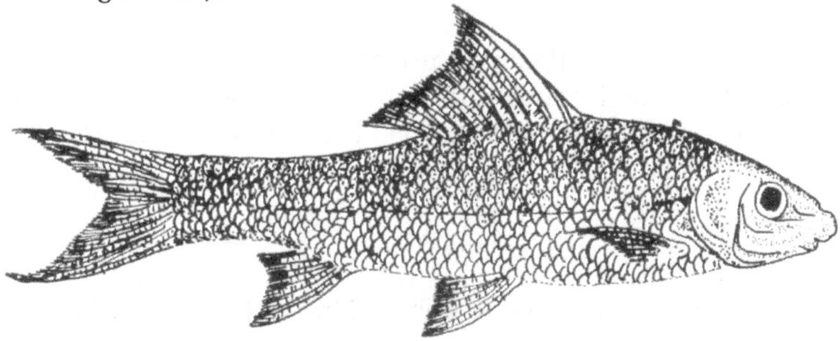

Figure 281: ***Labeo dero***

L. gonius

Local name is kursi and bata. It is present in whole northern India and upto Krishna in the south *(Figure 282)*.

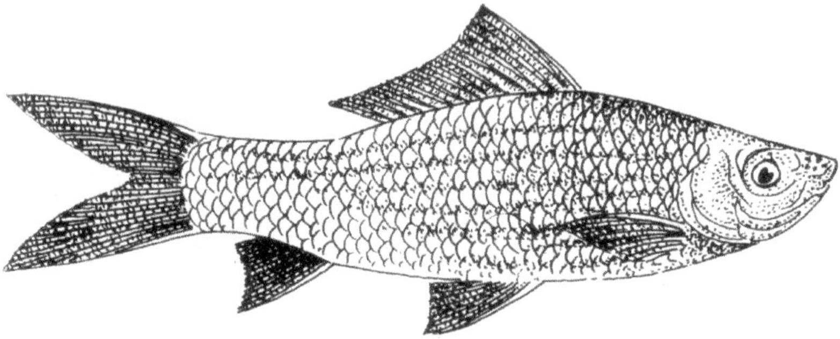

Figure 282: *Labeo gonius*

L. pangusia

Local name is rewa or loanee. It is found in Bihar, Bengal and Assam *(Figure 283)*.

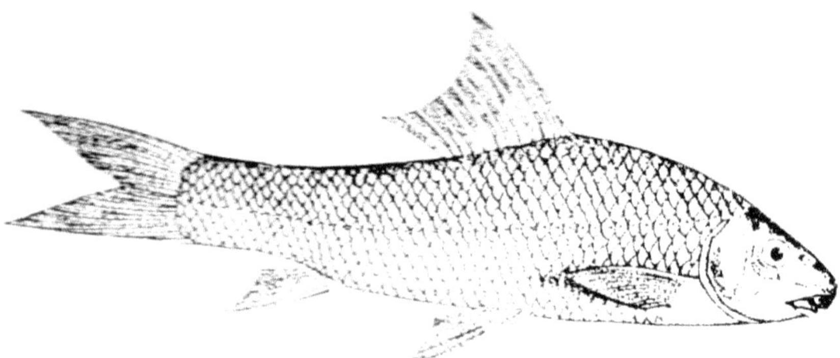

Figure 283: *Labeo pangusia*

Genus–*Cirrhinus*

Various species are as follows:

Cirrhinus mrigala

Popularly known as nain (Bengali–mrigala; Telugu–yerra mosu; Oriya–mirrgah, Punjabi–mori). Its pectoral, pelvic and anal fins are arrange in columns and tipped with black. A pair of small barbels are also present. It is found in the ponds and rivers of India. Its colour is silvery dark grey along the back and whitish below *(Figure 284)*.

Cirrhinus reba

It is commonly called raia and distributed throughout India. Silvery and scales are generally darkest at their edges *(Figure 285)*.

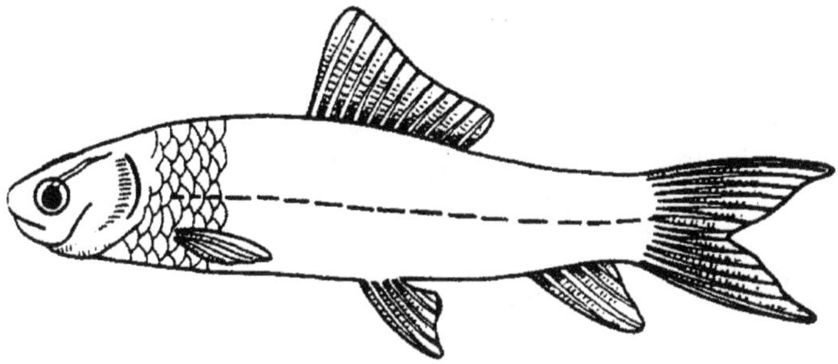

Figure 284: ***Cirrhinus mrigala***

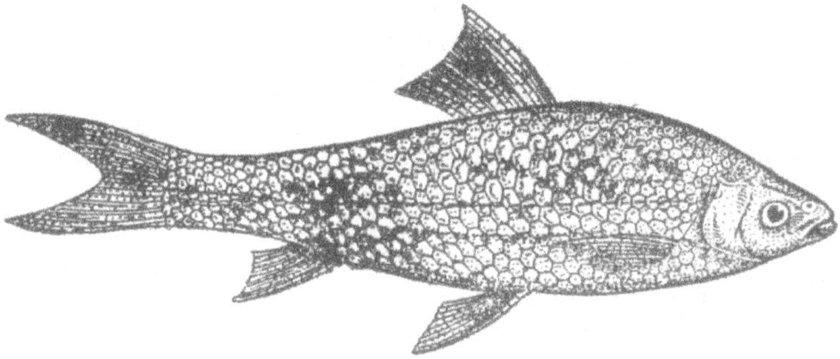

Figure 285: ***Cirrhinus reba***

Cirrhinus chaudhryi

Its local name is nain. It possesses two pairs of barbels and colour is silver with dark grey along the back. Pectoral, ventral and anal fins are white tinged with grey *(Figure 286).*

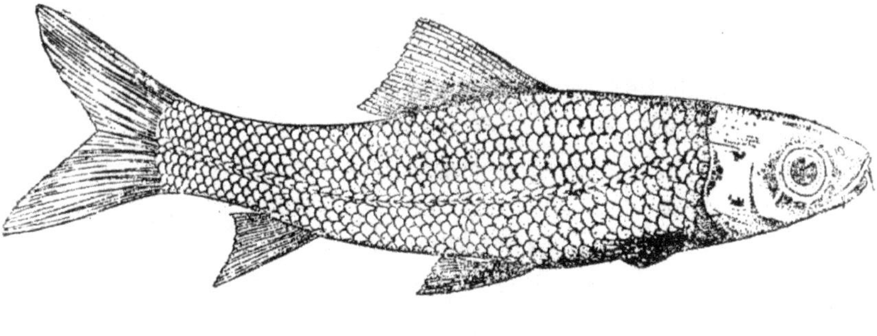

Figure 286: ***Cirrhinus chaudhryi***

Genus–*Catla*

Catla catla

Popularly known as catla (Bengali–katla; Kutch and Marathi–tambra; Malayalam–karakatla; Oriya–barkur; Punjabi–theimla; Uttar Pradesh and Bihar–bhakur; Assam–budhekera). It is the largest Indian carp found throughout India. It is one of the fastest growing fishes *(Figure 287)*.

Figure 287: ***Catla catla***

Tor/Mahseer

Known globally for game and table values. The important species are :-

Tor tor

Popularly known as mahasher or nahran (Bengali–tor, mahasol, putitor; Marathi–khadchi, mesta; Kannada–peruval; Malayalam–meruval; Uttar Pradesh and Bihar–mahasher). It grows to a large size of about 1.20 metre and ranks very high as food fish. Body is covered with large cycloid scales and its scale is rated as the largest among the modern teleosts. Its fins are yellow with red margins. It is also a good game fish *(Figure 288)*. It has been found in the foothills of Himalaya, Assam, Bihar, Madhya Pradesh and northern parts of Bengal.

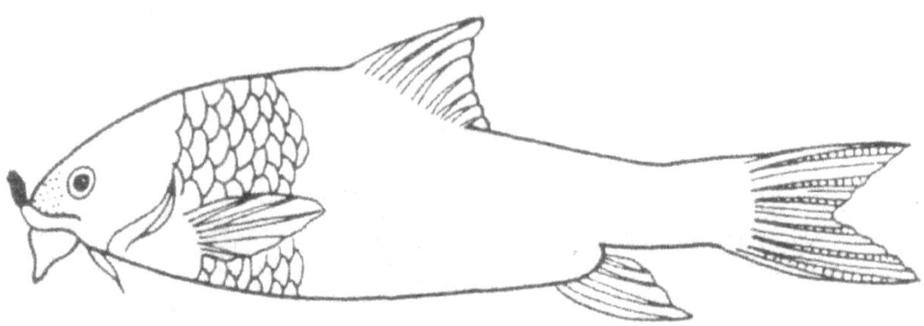

Figure 288: ***Tor tor***

T. putitora

In India, it has been found all along the Himalaya. It has a pointed head and lateral line. Scales are 23-28 *(Figure 289)*.

T. khudree

The so-called Deccan mahseer has been found in Peninsular India especially Karnataka, Kerala, Maharashtra's hill streams. Fins are bluish-grey. There are 24-26 scales along the lateral line *(Figure 290)*.

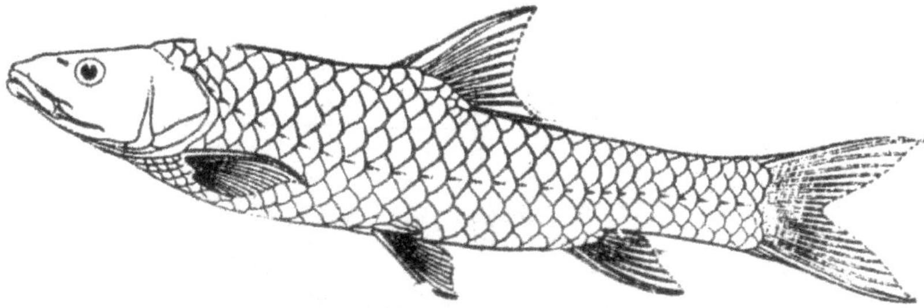

Figure 289: ***Tor putitora***

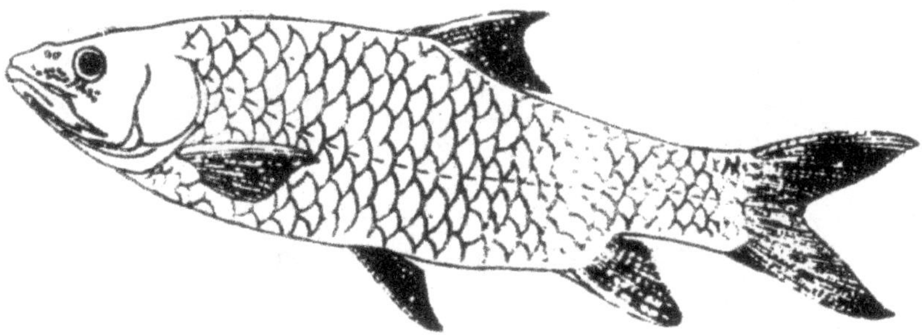

Figure 290: ***Tor khudree***

T. progenius

In India it has been found in Arunachal Pradesh, Assam, Naga hills and Nagaland. The number of scales along lateral line are 27-31.

Wallago attu

Popular name is boalee (Bengali–boyali; Marathi–shivada; Telugu–valuga; Tamil–valai; Kannada–bahle; malayalam–attuvalai; Punjabi–mully; Uttar Pradesh and Bihar–parhin. It is commonly known as freshwater shark. It has two pairs of barbels (maxillary and mandibular), one long pair on the upper jaw (maxillary) and a short mandibular on the lower (*Figure 291*).

Schizothorax (Snow Trout)

Schizothorax richardsonii

It is an indigenous snow trout of Kashmir and other Himalayan streams. Its body is streamlined with inferior mouth. A sucker on the lower lips is formed of raised papillar and possess 97 scales on lateral line. Possess two pairs of barbels. The length of head is 4.4 in standard length of fish. It is one of the important game fish. It is found in snowfed streams of Assam and Eastern Himalayas through Bhutan, Sikkim and Nepal to Punjab, Kashmir *(Figure 292)*.

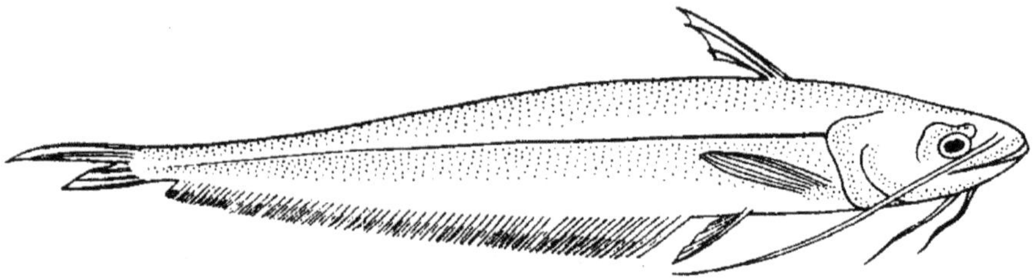

Figure 291: ***Wallago attu***

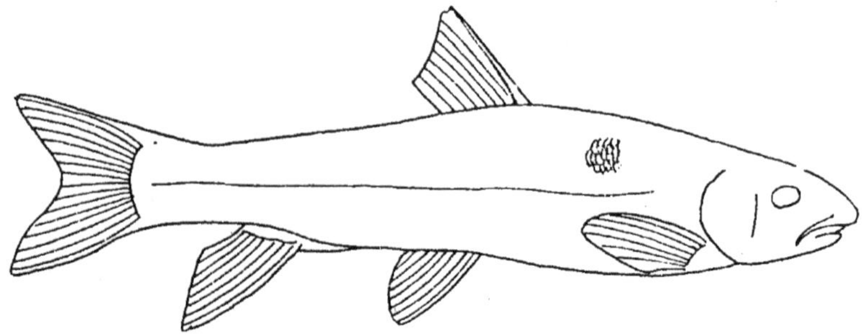

Figure 292: ***Schizothorax richardsonii***

S. kumaonensis

It has been recently described from Kumaon region. This species is diagnosed by a suctorial mouth, 97-98 scales in lateral line. The length of head is 5.4 in standard length.

Schizothoraichthys

In India, its species are endemic to Kashmir.

Schizothoraichthys esocinus

This species occurs in Kashmir and Ladakh (*Figure 293*).

S. progastus

It is also regarded as a true snow trout. It occurs from head-waters of Ganga to Sadiya in Upper Assam through Darjeeling and is common at Haridwar.

S. planifrons

In India–Kashmir.

S. micropogon

India–Kashmir.

S. longipinnis

India–Kashmir.

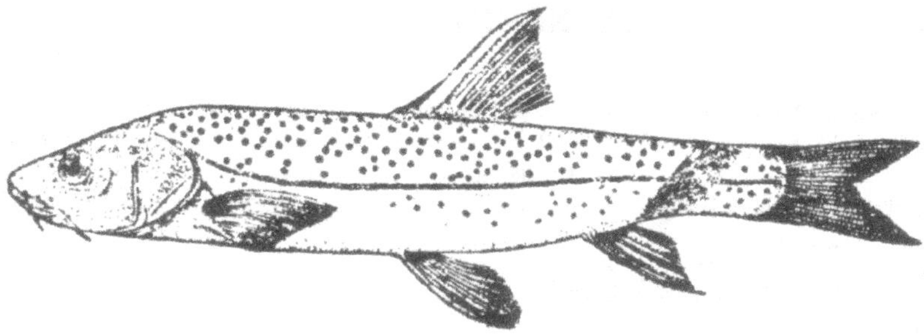

Figure 293: ***Schizothoraichthys esocinus***

S. hugelii

India–Kashmir.

S. curvifrons

India–Kashmir.

Ctenopharyngodon

C. idella (Grass Carp)

It is an exotic carp and natural inhabitant of China and USSR. This species was introduced in India in the ponds of Central Inland Fisheries Research Station (CIFR), Cuttack, Orissa in 1959. It has a fast growing rate *(Figure 294).*

Figure 294: ***Ctenopharyngodon idella***

Cyprinus carpio (*Figure* 235)

This common crop is distributed all over the world and its three known varieties are:

Cyprinus carpio communis (Scale Carp)

Its body is fully covered by regularly arranged rows of scales and supposed to be the original form.

Cyprinus carpio specularis (Mirror Carp)

This carp is now well established in Himachal Pradesh, Punjab, Jammu & Kashmir, Madhya Pradesh, Rajasthan, Delhi, Bihar, Manipur, Sikkim. It is an ideal species and for introduction into hilly regions. Body is covered unevenly with a few large and bright scales. A large area of body is without scales (*Figure 123*).

***Cyprinus carpio nudus* (Leather Carp)**

Its body is without scales except for a single row of degenerate ones along base of dorsal fin occasionally extending as far tail or even from head to caudal fin. Base of other fins may also be covered with small scales.

Hypophthalamichythys (Silver Carp)

H. molitrix

It is a native of China and introduced in India in 1959. It grows very fast. Its abdominal edge is keeled (*Figure 295*).

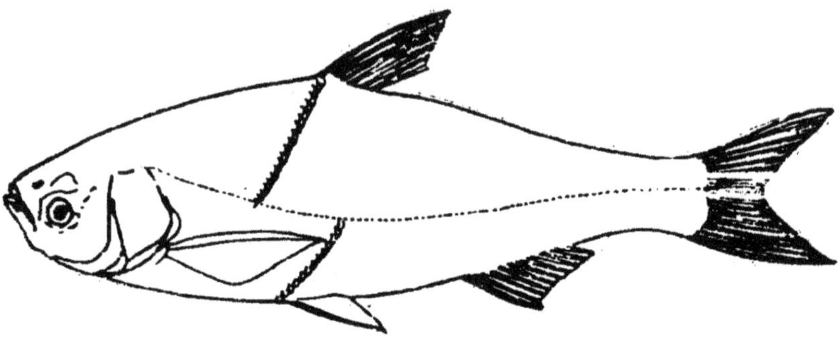

Figure 295: ***Hypophthalamichythys molitrix***

H. nobilis

It has been recently introduced. Only part of the abdomen is keeled.

Among cat fishes, the common fishes of different genera are:

Genus–*Mystus*

Body is short, moderately elongated, abdomen is rounded. Head is of moderate size and compressed. Snout is rounded or obtuse and lips are thin. Four pairs of barbels, one each of maxillary and nasal and two of mandibular, generally longer than head. Some of its edible species are:

M. seenghala

Popular name is tengra or ari (Bengali–air; Marathi–singala; Telugu–multijellah; Tamil–cumbookelutti; Kannada–shedi; Malayalam–keratta; Punjabi–shingaree; Uttar Pradesh and Bihar–dariai tengar). It has a well developed adipose fin with a circular black spot. It is widely distributed in India, being found in the Yamuna, Ganges, Deccan and Assam rivers. (*Figure 296*).

M. vittatus

Its common name is tengara. It occurs throughout India but is more common in Uttar Pradesh and Tamil Nadu (*Figure 297*).

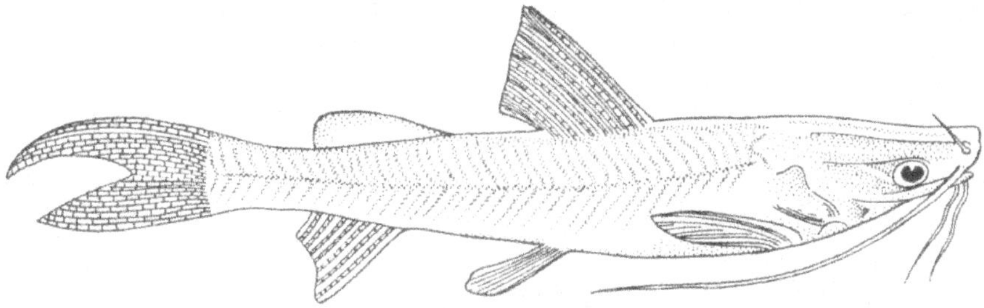

Figure 296: ***Mystus seenghala***

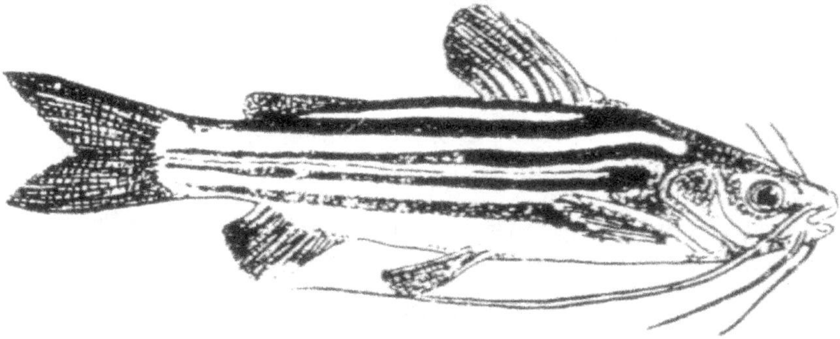

Figure 297: ***Mystus vittatus***

M. aor

It is widely distributed in India and grows to a size about a metre long. The maxillary barbels are exceptionally long, reaching nearly base of caudal fin or even beyond it.

M. cavasius

Its common name is sutahawa tengara. It is common in freshwaters of all over India. Its size is about half a metre long (*Figure 298*).

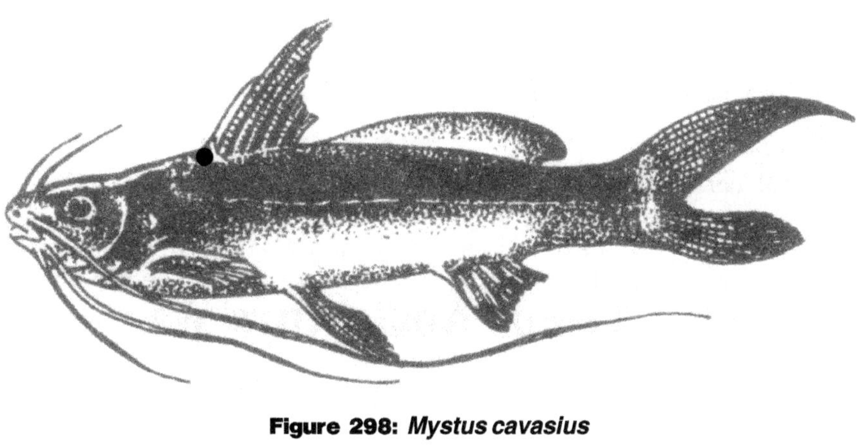

Figure 298: ***Mystus cavasius***

M. bleekeri

In India, it is generally confined to north India; the southmost limit to the Mahanadi head waters, commonly found in West Bengal only (*Figure 299*).

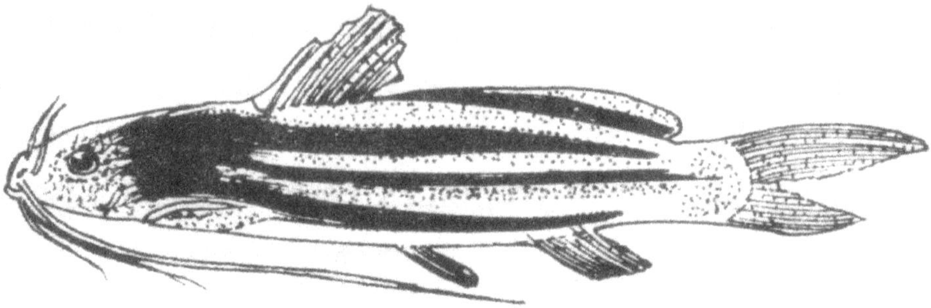

Figure 299: *Mystus bleekeri*

M. tengara

Its local name is tengana and is distributed in northern part of India and Madhya Pradesh (*Figure 300*).

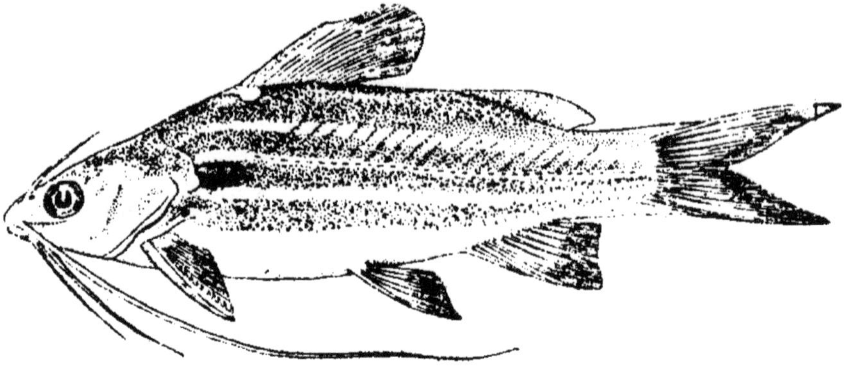

Figure 300: *Mystus tengara*

Genus–*Ompok*

The important species are:

O. bimaculatus

Popular name is puffta (Telugu–duka demu; Tamil–chotawahla; Kannada–godla; Malayalam–wahlah; Punjabi–pufta; Uttar Pradesh and Bihar–jalkapoor). It is an excellent food fish. Its dorsal fin is very short, body is laterally compressed and head is depressed. A black spot is present on the shoulder and another at the tail. The anal fin is covered with a leathery flap on each side (*Figure 301*).

O. pabo

In India, it is present in Brahamputra, Ganga, Yamuna river systems (*Figure 302*).

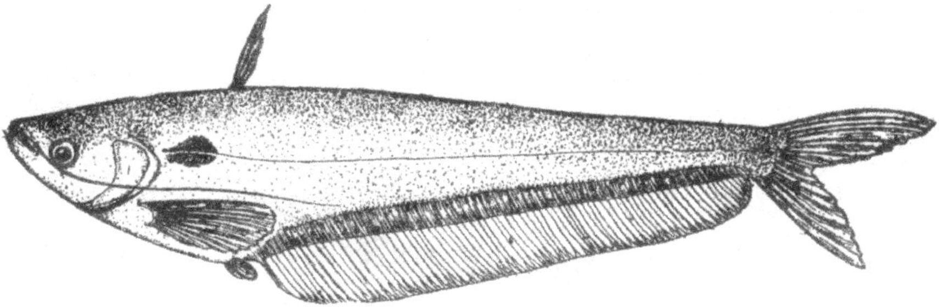

Figure 301: ***Ompok bimaculatus***

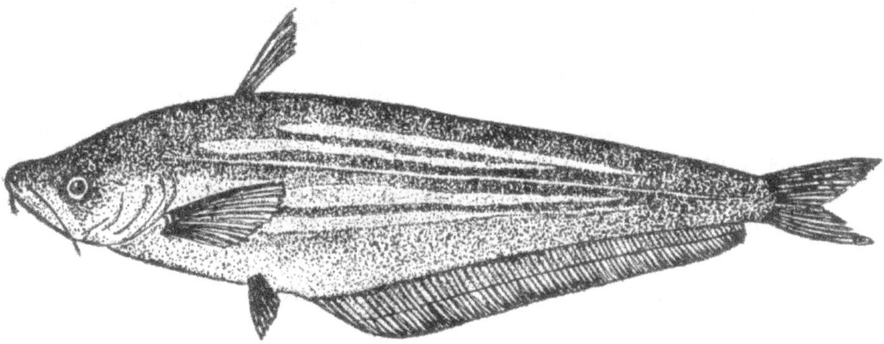

Figure 302: ***Ompok pabo***

O. pabda

It is present in Brahamputra and Ganga river systems (*Figure 303*).

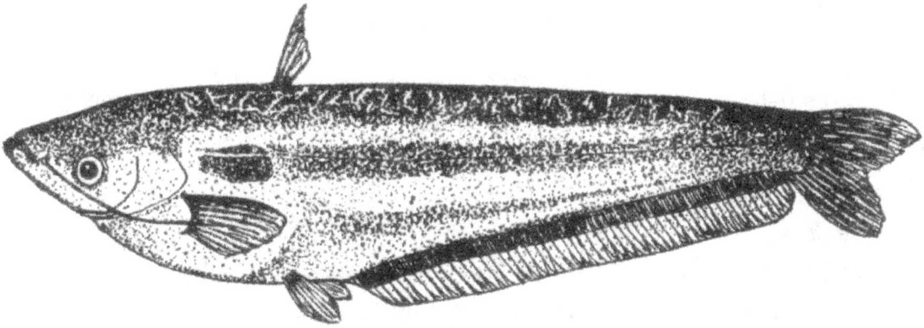

Figure 303: ***Ompok pabda***

Genus–*Pangasius*

P. pangasius

Popular name is pariasi (Bengali–pungar; Telugu–choluvajella; Tamil–kovailoolakeluthai; Uttar Pradesh and Bihar–payas). It is present both in freshwater and estuarian. Its body is elongated and compressed with a short anal fin. First dorsal fin is short with a spine provided with poison gland (*Figure 304*).

Figure 304: ***Pangasius pangasius***

Genus–*Heteropneustes*

H. fossilis

Popular name is singhi (Marathi–bitchuka mach; Telugu–mapujulla; Malayalam–kahrimeen; Punjabi–noorie; Bengali, Uttar Pradesh and Bihar–singhi). This fish has eight barbels. Dorsal fin is short without spine, anal fin is elongated and the pectoral spine of this fish are poisonous. This is a tasty and nourishing food fish. Its anal fin is separated from caudal by a notch (*Figure 305*).

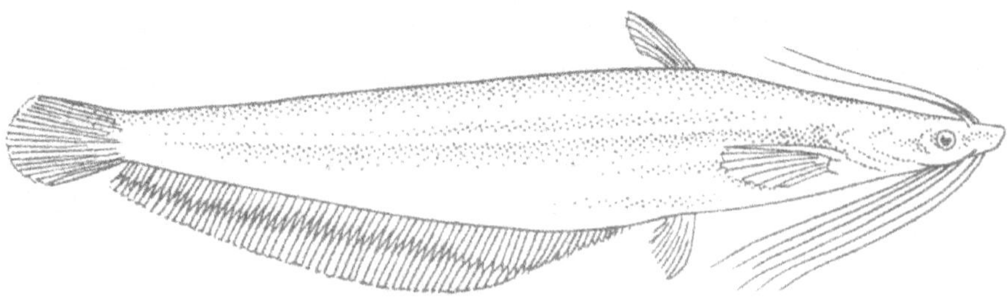

Figure 305: ***Heteropneustes fossilis***

H. microps

In this fish anal fin is confluent with caudal fin. In India, it has been found in Bihar and Uttar Pradesh.

Genus–*Clarias*

C. batrachus

Popularly known as mangri (Bengali–mah-gur; Telugu–marpoo; Tamil–masarai; Malayalam–yerivahlay; Punjabi–kug-ga; Uttar Pradesh and Bihar–magur). It is supposed to be a poorman's fish and is found in dirty, muddy bottom of fresh and brackish waters. It is a delicious fish and fetches more price than carps. It is found in fresh and brackish water of India (*Figure 306*).

Body is Eel-like and Minute Scales are Embedded in the Skin (Order: Arguilliformes)

Genus–*Anguilla*

A. bengalensis

Popularly known as ban or van (Marathi–ahir; Telugu–malungula; Tamil–serumpambu, Malayalam–maningal). It is a well known freshwater eel and looks

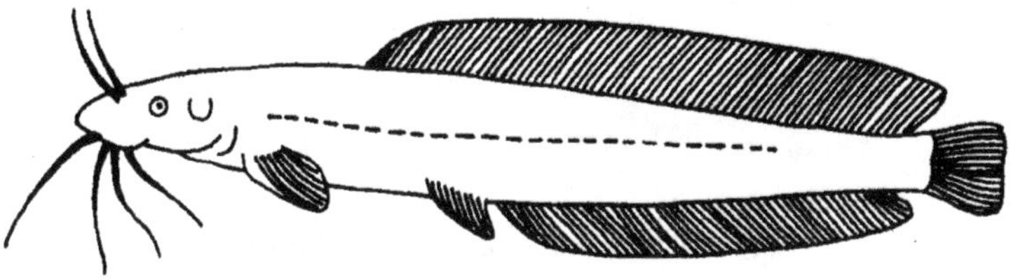

Figure 306: ***Clarias batrachus***

like a snake, but it is a poorman's food fish. Body is scaleless with long dorsal fin and missing ventral and pectoral fins (*Figure 307*).

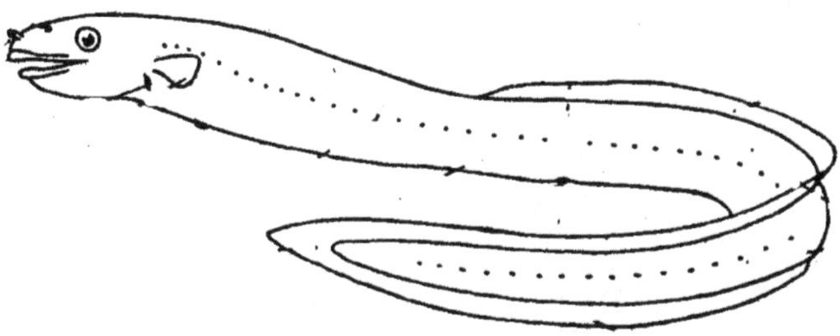

Figure 307: ***Anguilla bengalensis***

Body is Elongated Eel-like with Longer Lower Jaw.

Some free spines are present in front of dorsal fin (order: mastacembeliformes). The important edible fishes are:

Genus–*Mastacembalus*

M. armatus

It is present throughout India and is popularly known as baan. It prefers to live in mud. Its flesh is very tasty from consumption point of view (*Figure* 222).

M. pancalus

Popularly known as malga (Uttar Pradesh and Bihar–patya). It is eel-like with yellow dorsal, pectoral, caudal and anal fins with black spots which are arranged with transverse bars on the anal fin. Its flesh is also very tasty and it is a fish of poors (*Figure 308*).

When the Fins are with Spines, Scales are of Ctenoid Type, Ventral Fins are Thoracic, may be Marine or Freshwater (Order: Perciformes)

Few fishes of this order are freshwater and these are excellent food fishes. These are characterised by fins with spines, two dorsal fins and ventral thoracic fin. The important food fishes of this order are:

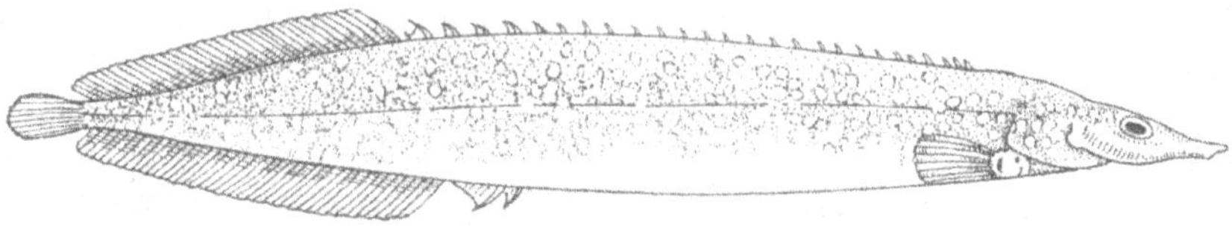

Figure 308: ***Mastacembalus pancalus***

Genus–*Sciaena*

S. coitor

Popularly known as patharchatti (Bihar–bhola; Tamil–vella kateelee; Malayalam–kora). Its flesh is very tasty. It touches the sea in one season and Himalayas in the other (*Figure 309*).

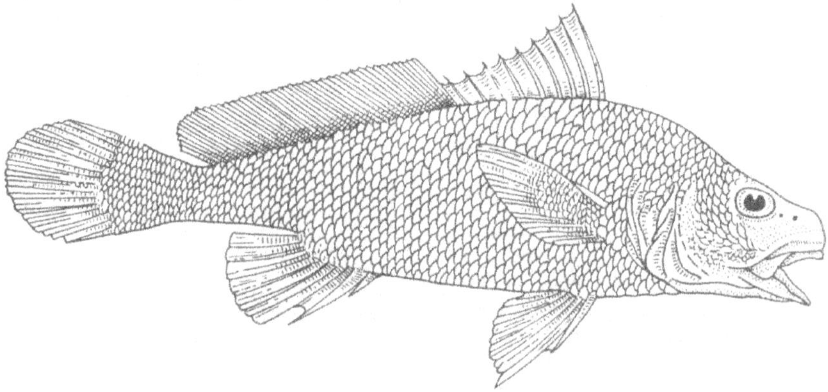

Figure 309: ***Sciaena coitor***

Genus–*Anabas*

A. testudineus

Popularly known as koi or kobhai (Uttar Pradesh and Bihar–Sumha). It is a highly nutritive and costly fish. Its body is oblong, almost cigar-shaped with wide head. It has a pleasing flavour and is very popular as food. It has special labryinthine accessory breathing organs which enable it to move on land for some distance (*Figure 310*).

When the Fins are Without Spines, Fish is Snake Headed and Head is Covered with Cyloid Scales (Order: Ophiocephaliformes)

The important edible fishes are:

Genus–*Channa*

C. punctatus

Popularly known as phool-dhok (Bengali–taki; Telugu–mitta; Tamil–korava; Malayalam–kayi chil; Punjabi–dulloonga; Uttar Pradesh and Bihar–girai). It is

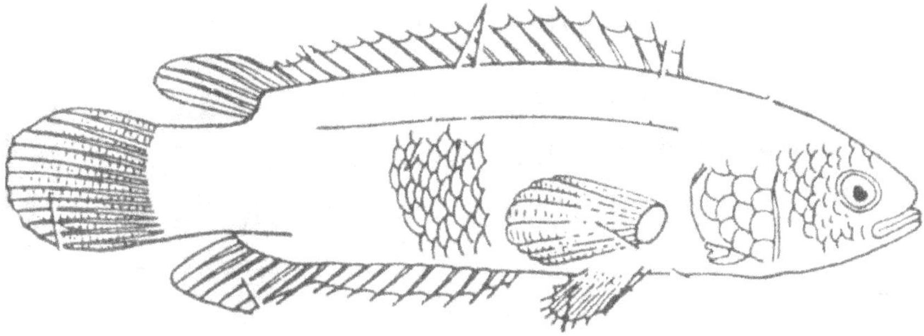

Figure 310: ***Anabas testudineus***

medium sized murrel and is esteemed as food. Its dorsal and anal fins are elongated, caudal fin is separate and rounded (*Figure 311*).

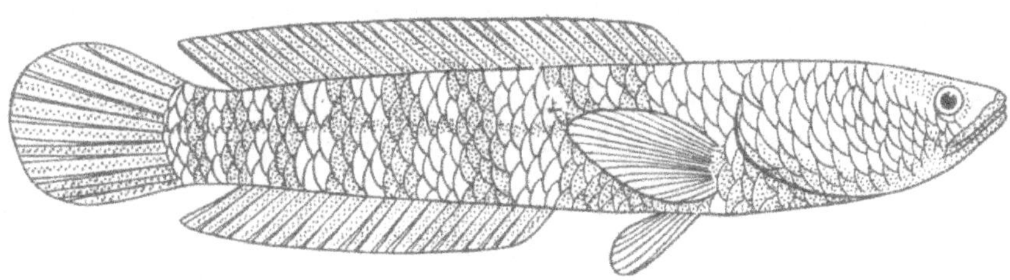

Figure 311: ***Channa punctatus***

C. striatus

Commonly known as morrul (Bengali–shol; Marathi–sohr; Telugu–korrameen; Tamil–vishal; Malayalam–kannam; Punjabi–sowl; Uttar Pradesh and Bihar–saure). This fish prefers to live in stagnant muddy waters. It is very economical and more palatable than major carps. It is a typical "live fish" with soft flesh devoid of fats and is considered to have medicinal value and provides nutritious food to the sick (*Figure 312*).

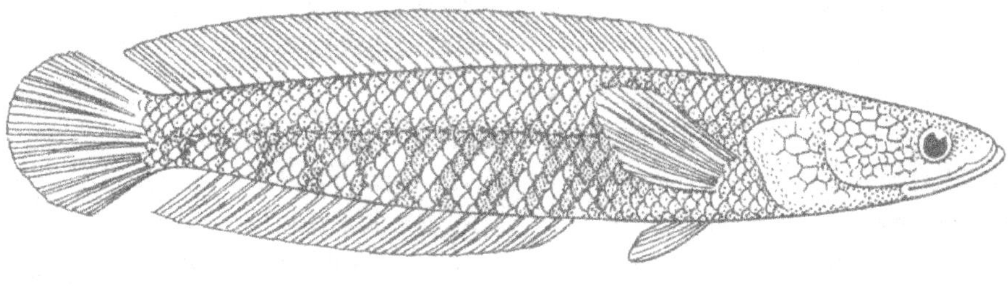

Figure 312: ***Channa striatus***

C. maralius

Commonly named as pumuri (Bengali–sal; Telugu–pool-a-amalle; Tamil–aviri; Kannada–madinji; Malayalam–bral; Punjabi–kubrar or sawl; Uttar Pradesh and

Bihar–saur). It is largest of murrels in India. It is valued as an esteemed food (*Figure 313*).

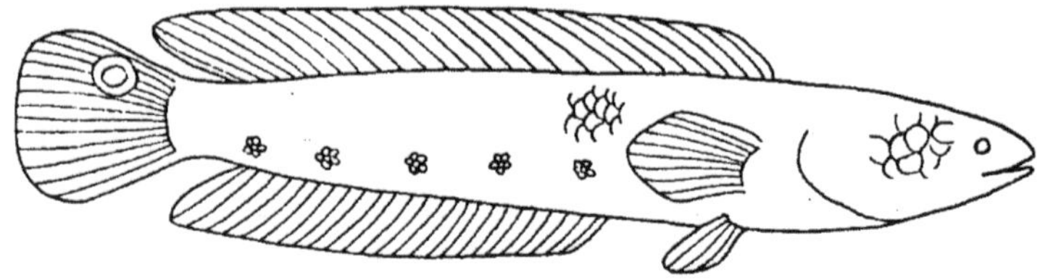

Figure 313: ***Channa maralius***

Fishes Resemble Perches, but the Ventrals are Abdominal or Subabdominal in Position, Head is Suppressed Anteriorly, Body may be Oblong, Subcylinderical or Compressed with Two Dorsal Fins and Lack Lateral Line (Order: mugiliformes)

Majority of species are marine and estuarine but a few are cultivated in ponds and tanks and used in freshwater fishery as food fishes.

Genus–*Mugil*

M. corsula

Popularly known as grey mullet (Uttar Pradesh and Bihar–hunra). Eyes are without adipose lids but bulged on the dorsal side. Dorsal fins are two in number and upper lip is thick and notched. It is known to be an excellent food fish (*Figure 314*).

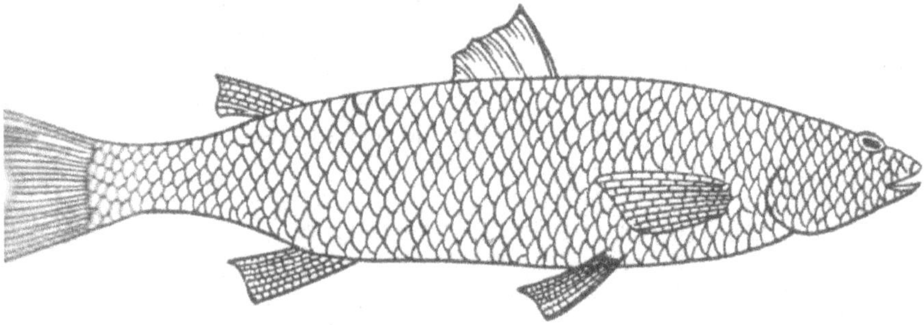

Figure 314: ***Mugil corsula***

M. cephalus

Popularly known as grey mullet (Bengali–ain; Mumbai–boe, Telugu–kathiparego, Tamil–madarai, Kannada–mala, Malayalam–tirutha). It is cultivated in the tanks of Tamil Nadu and Kerala states of India. (*Figure 315*).

When the Jaws are Beak-like, Anal and Dorsal Fins are Posteriorly Situated, Scales are Thin and Cycloid, Generally Freshwater Inhabitant (Order: Beloniformes)

The fishes of this order are excellent as food fishes. Their body is prominent beak-like. The dorsal and anal fins are posteriorly situated and opposite each other

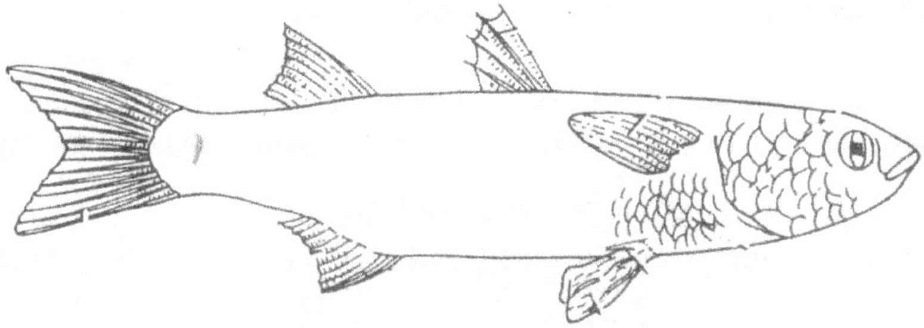

Figure 315: ***Mugil cephalus***

ventrals are abdominal and pectorals are inserted high ups. The common food fishes are:

Belone cancila or *Xenentodon cancila*

Popularly known as kowa (Bengali–kankle; Telugu–vaddla-mukku; Tamil–kokkumeen; Kannad–konti; Malayalam–kolan; Uttar Pradesh and Bihar–kauwa). It is esteemed as food fish. The jaws are produced in beak and lower jaw is slightly longer. Caudal fin is slightly emarginate and dorsal fin commences opposite the anal (*Figure 316*).

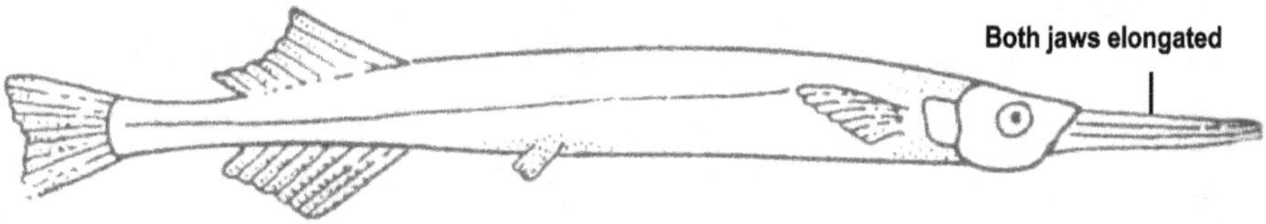

Figure 316: ***Xenentodon cancila***

Hemiramphus gorakhpurensis

Popular name is half-beaks (Telugu–gonia; Tamil–kozhutha; Kannada–kande; Malayalam–koyala; Uttar Pradesh and Bihar–kauwa-ka-bachcha). In this fish only lower jaw is elongated into beak and prominent. It affords local fishery in eastern Uttar Pradesh and is an important food fish (*Figure 317*).

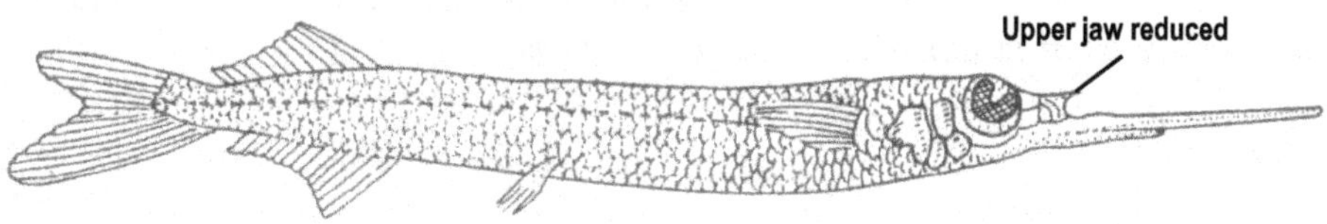

Figure 317: ***Hemiramphus gorakhpurensis***

Determination of Freshness of Common Edible Species of Domestic Use

Fish removed from water cannot survive and dies quickly of suffocation. It remains in good condition for a few hours after which decomposition begins due to bacterial action, enzymatic action or chemical action. Fish, after its death, remains soft and fresh for 5-7 hours, and then its muscles start becoming stiff. This stiffness or rigidity of the muscles is called *'rigor mortis'*. It lasts for several hours during which the fish is regarded 'fresh and fit' for human consumption. On the basis of physical appearance we can diagnose the freshness of fish. Some of the common features of fresh fish are as follows:

1. Skin and colour are bright and not dull, spotted, slimy, pale or bleached.
2. In normal looking fish, eyes are full, prominent, glistening and not sunk and opaque.
3. Scales adhere together and are not loose.
4. The gill lamellae appear bright red and not yellow, brown or grey.
5. Vent is normal and not protruding.
6. The body is firm and stiff and not soft or flabby. A finger pressed against the body does not leave a permanent depression of the finger print but the latter immediately recovers.
7. The slime of the body smells fish like but and not foul like.
8. The body sinks in water and not floats.

During *rigor mortis*, the body loses flexibility and attains stiffness. This stiffness progresses in cephalo-caudal direction, the mouth opens, gill covers spreads out and body axis arches. For a certain duration, the fish body remains in *rigor mortis*. After this period is over the fish again becomes soft, then limp and finally flabby. From death upto *rigor mortis* stage the fish is good, fresh and edible. After *rigor mortis*, the fish becomes rotten, inedible from the point of human consumption. *Rigor mortis* is the index of freshness and high quality of fish, and is taken as a test for determining the freshness of fish.

Rigor mortis lasts for 30-120 hours. The pH of a freshly killed fish ranges between 7–7.35 and after death drops to 6.2–6.5

Fish preserved prior to *rigor mortis* is inferior in taste and quality as compared to fish preserved after the setting-in of *rigor mortis*.

//••••••••••//

References

Arrignon, Jacques (1999). Management of Freshwater Fisheries. Oxford and IBH Publishing Corporation Pvt. Ltd., 66, Janpath, New Delhi, p. 582.

Bailey, R.M. (1960). Pisces (Zoology) pp. 242-243 in McGraw-Hill Encyclopedia of Science and Technology, vol. 10, New York, McGraw-Hill.

Brandt, A.V. (1984). Fish catching methods of the world. Fishing News (Books) Ltd., London: 432 p.

Cohen, Daniel M (1970). Readings in Ichthyology edited by Milton S. Love and Gregor M. Cailliet.-Prentice-Hall of India Private Limited, New Delhi-110001, 1979: p. 24-27.

Day (1878). The fishes of India: being a natural history of fishes known to inhabit the seas and freshwaters of India, Burma and Ceylon. Reproduced in 1858: William Dawon and Sons, 778 pp.

Eddy, Samuel and James C. Underhill (1969). How to know the freshwater fishes. The Pictured Key Nature Series N.M.C. Brown Company Publishers, Dubuque, Iowa.

FAO (1998). FAO Yearbook: Fishery Statistics. Capture Production, 1996, vol. 82, FAO, Rome.

Goodrich, Edwins (1984). Studies on the structure and development of vertebrates, vol. 1. International Books and Periodic Supply Service, 24 B/5, Deshbandhu Gupta Road, Karol Bagh, New Delhi, p. 97.

Hameed, M. Shahul and M.R. Boopendranath (2000). Modern Fishing Technology. Daya Publishing House, Delhi–110035.

Hangay, George and Dinglay Michael (1985). Biological Museum Methods, Vol. 1, Academic Press Australia, p. 18.

Harder, Wilhelm (1975). Anatomy of Fishes. E. Schweizerbartsche Verlagsbuch handlung (Nagele U. Obermiller) Stuttgart.

Hora (1951a). Knowledge of the ancient Hindus concerning fish and fisheries of India. Matsyavinoda or a chapter on Angling in the Monosoltara by King Somesvara (1127 AD). J. Asiat. Soc. Bengal 17 (2): 145-69.

Hora (1956). Fish paintings of 3rd Millennium B.C. from Nal (Baluchistan) and their zoogeographic significance. Mem. Indian Mus. 14 (2): 78-84.

Jayaram, K.C. (1999). The Freshwater Fishes of the Indian Region. Narendra Publishing House, p. XXIII–Introduction.

Jayaram, K.C. (2002). Fundamentals of Fish Taxonomy. Narendra Publishing House, Delhi, p. 2-60.

Jhingran, V.G. (1991). Fish and Fisheries of India. Published by Hindustan Publishing Corporation, 3rd edition, page no. 3.

Jhingran, V.G. and K.L. Sehgal (1978). Cold Water Fisheries of India. Printed in India by The Perfect Printers, 73/1/2 Hospital Road, Barrackpore, West Bengal.

Kapoor, B.G. and Bhavna Khanna (2004). Ichythyology Handbook. Narosa Publishing House, New Delhi, p. 47.

Khanna, S.S. and H.R. Singh (2003). A Textbook of Fish Biology and Fisheries. Narendra Publishing House. Publisher and Distributors, 1417, Krishan Dutt Street, Maliwara, Delhi–110006.

Laevastu, T. (1965). Manual of methods in fisheries biology. Section IV, Fascicule 9, FAO Manuals in Fisheries Science Number 1, page 27.

Lagler, Karf F., John E. Bardech, Robert R. Miller, Dora R. May Passino (2003). Ichthylogy. John Wiley & Sons (ASIA) Pte. Ltd., 2 Clementi Loop # 02-01, Singapore.

Lindberg, G.U. (1976). Fishes of the World. A Halsted Press Book. John Wiley & Sons, New York-Toronto. Israel Program for Scientific Translations.

Marshall, N.B. (1965). The Life of Fishes. London, Weidenfeld and Necolson, 402 pp.

Moyle, Peter B. (1993). Fish: An Enthusiast's Guide. University of California Press, Berkeley, Los Angeles, California. Printed in United States of America.

Munshi, J.S. Dutta and M.P. Srivastava (2002). Natural History of Fishes and Systematics of Freshwater Fishes of India. Narendra Publishing House, Delhi (110006).

Myers, G.S. (1958). Trends in the evolution of teleostean fishes. Stanford Ichthyological Bulletin, vol. 7, pp. 27-30.

Nikolsky, G.V. (1999). Ecology of Fishes. Published by Allied Scientific Publishers, 4E 177 J.N. Vyas Nagar, Bikaner 334003, India.

Norman, J.R. (1963). A History of Fishes. 2nd edition of P.H. Greenwod, New York, Hill and Wang xxxi, 398 pp.

Norman, J.R. (2002). History of Fishes. Published by Asiatic Publishing House, 181/4, Exdtn. Street No. 6, Laxmi Nagar, Delhi–110092.

Schultz, L.P. (1965). Fishes and how they live. Chapt. 1 in Wondrous World of Fishes. Washington DC, National Geographic Society, 367 pp.

Schultz, L.P. and E.M. Stern (1948). The Ways of Fishes. New York, Van Nostrand xii and 264 pp.

Srivastava, C.B.L. (2000). A Text Book of Fishery Science and Indian Fisheries, Kitab Mehal, 22-A, S.N. Marg, Allahabad.

Srivastava, Gopalji (1998). Fishes of UP and Bihar. Published by Vishwavidyalaya Prakashan Chowk, Varanasi (India).

Waller, R.A. and W.N. Eschmeyer (1965). A method of preserving colour in biological specimens. Bioscience p. 36.

Index

www.ingramcontent.com/pod-product-compliance
Lightning Source LLC
Chambersburg PA
CBHW071245230926
51849CB00001B/233
* 9 7 8 9 3 5 1 3 0 9 8 2 6 *